PIONEERS *of* MICROBIOLOGY
and the
NOBEL PRIZE

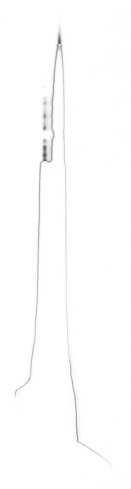

PIONEERS *of*
MICROBIOLOGY
and the
NOBEL PRIZE

Ulf Lagerkvist
Goteborg University, Sweden

World Scientific
New Jersey • London • Singapore • Hong Kong

Published by

World Scientific Publishing Co. Pte. Ltd.
5 Toh Tuck Link, Singapore 596224
USA office: Suite 202, 1060 Main Street, River Edge, NJ 07661
UK office: 57 Shelton Street, Covent Garden, London WC2H 9HE

Library of Congress Cataloging-in-Publication Data
Lagerkvist, Ulf.
  Pioneers of microbiology and the Nobel prize / Ulf Lagerkvist.
    p. cm.
  Includes bibliographical references and index.
  ISBN 981238233X (alk. paper) -- ISBN 9812382348 (pbk. : alk. paper)
    1. Microbiology--History--19th century. 2. Microbiologists--History--19th century.
  3. Nobel Prizes--History--19th century. I. Title.

QR21 .L34 2003
579'.09034--dc21   2002033194

British Library Cataloguing-in-Publication Data
A catalogue record for this book is available from the British Library.

Cover images courtesy of Nobel Foundation, Bonnierförlagen and
Royal Swedish Academy of Sciences, Stockholm, Sweden.

Printed in Singapore by Mainland Press

Contents

Preface

This book is about four great pioneers of medical microbiology — Emil von Behring, Robert Koch, Paul Ehrlich and Elie Metchnikoff — and their scientific contributions. They were all awarded the Nobel Prize in physiology or medicine. Thus, it may be of interest to see what impression their discoveries have made on contemporary science and particularly on colleagues at the Karolinska Institutet in Stockholm, who had been charged with the responsibility of evaluating the candidates for the early medical Nobel prizes. To appreciate their achievements, it is necessary to give the reader a short account of the breakthrough of scientific medicine in the 19th century and in particular the advent of medical microbiology. The book is primarily intended for an audience of laymen and I have tried to avoid being too scientifically ambitious, without falling into the trap of presenting only entertaining chatter.

I am greatly indebted to the Nobel Assembly/Nobel Committee of the Karolinska Institutet for generously giving me access to their archive with information on Behring, Koch, Ehrlich and Metchnikoff.

Introduction

The teachings of Hippocrates dominated medical thinking from the 4th century B.C. to the beginning of the 19th century. They first spread to Rome, where the Greek physician Galen (A.D. 130–200) was active in all fields of medicine. His writings were studied by physicians also in the Arab world empire, where Hippocratic medicine found a refuge from the harsh cultural climate that prevailed in Western Europe during the Middle Ages. As Islamic medicine, it re-entered Southern Europe in the 10th century and established itself at the famous medical school of Salerno and the medical faculties of the universities of Bologna and Montpellier.

It was not the acute clinical perception of the Hippocratic physicians, which we nowadays admire so much, that gave classic medicine its amazing power of survival. Instead it was the doctrine of the four body humors. Their correct proportions determined the health of the individual and all diseases could ultimately be traced to a disturbance of the delicate balance between blood, phlegm, yellow and black bile. Hippocratic medicine was therefore rather indifferent to a correct diagnosis, which we consider to be all-important. In the final analysis, all diseases had the same origin, disturbances of the body fluids, and the treatment was always the same — different kinds of blood-letting, emetics and laxatives. It was all aimed at restoring the correct balance between the body humors.

Not even during the renaissance and the scientific revolution of the 17th century, when Andreas Vesalius (1514–1564) rejuvenated anatomy and William Harvey (1578–1657) laid the foundation of modern physiology with his discovery of the circulation of the blood, was there any change in the uncritical reliance of clinical medicine on Hippocrates and Galen. It was not until the beginning of the 19th century that scientifically based medicine started to emerge not only in the laboratories but also to some extent on the clinics. Towards the middle of the century, medical thinking becomes more and more centered around the structural element that seemed to be the ultimate building stone of all living organisms — the cell.

THE ERA OF THE CELL

The Sites of Diseases

The word *pathology*, like so many other terms in medicine, is Greek and means the science or study of suffering. Taken literally this would imply a science that goes back as far as the human race itself and even if interpreted as "the study of diseases" its origin is veiled in obscurity. On the other hand, it is easy to trace pathology in the more limited sense of pathological anatomy, i.e. the anatomical changes in organs and tissues caused by disease, to one dominating figure, the Italian anatomist Giovanni Battista Morgagni (1682–1771).

Morgagni was born in Forlì, a small town in northern Italy, and studied medicine in Bologna where he graduated in 1701. His teacher in anatomy was Antonio Valsalva, famous for his studies of the anatomy of the ear. Valsalva in turn had been a disciple of Marcello Malpighi, the founder of microscopic anatomy, illustrating vividly the dominating position of the Italian medical schools at the time. After a few years as a medical practitioner in Venice and later in Forlì, Morgagni was appointed to the chair of anatomy in Padua in 1715 and he held that position until his death at the age of 89. (There was no mandatory retirement age for professors in 18th century Italy!) Morgagni made a number of valuable contributions to anatomy, but his most important work by far was the monumental *De Sedibus et Causis Morborum per Anatomen Indagatis* ("*On the Sites and Causes of Diseases as Disclosed by Anatomical Dissections*").

He can hardly be accused of having rushed into print with a half-baked product; his *De Sedibus* was not published until 1761 when he was 79 years old. It has the somewhat original form of 70 letters written to a young colleague whose identity was never revealed. Possibly he is just the kind of literary mystification that the 18th century loved. What is so remarkable about Morgagni's book is that here, for the first time, disease is thought to be caused by processes in the organs of the body, not by disturbances in the balance of the body fluids as Hippocrates believed. Systematically, Morgagni goes through the pathology of the different organs and tries to relate what he finds at the autopsy to the clinical picture. Among his findings should be mentioned his description of the aortal aneurysm and his observation that stroke, with lesions in one hemisphere, leads to a hemiplegia involving the other side of the body.

The 18th century loved everything in nature that was curious and grotesque. The curiosity cabinets that adorned the mansions of the rich and powerful were very typical of this period. When Morgagni entered the medical scene, pathology was nothing more than a curiosity cabinet. The interest of such predecessors as the Swiss physician Théophile Bonet had been focused almost entirely on such matters. With *De Sedibus*, Morgagni single-handedly created a new branch of medicine, even if his colleagues did not fully appreciate the true importance of

Giovanni Battista Morgagni (1682–1771).
Courtesy of Bonnierförlagen, Stockholm, Sweden.

his life's work. Referring to the fanciful theories of contemporary medicine, Morgagni used to say that those who had performed many autopsies had at least learned to doubt, in contrast to the cocksureness of those who paid no attention to the anatomical changes in the dead body. The future would completely vindicate him and the remarkable development of pathology in the 19th century built on the foundation that he had laid.

Vitalism and Natural Philosophy

Of all the fanciful theories that flourished in medicine in the 17th and 18th centuries, vitalism was by far the most enduring. An early advocate of vitalism was the French physician Théophile de Bordeu (1722–1776), who spent most of his career in Paris, but had been educated in Montpellier which would become one of the strongholds of vitalism. Bordeu was above all fascinated by the glands whose function he considered to depend on a mystical vital force, hence the name vitalism. This somewhat obscure philosophy was of course in marked contrast to the ideas of the iatrophysicists or iatromechanics, as they were also called, with their simple mechanical explanations of body functions. To give an example, they saw the glands as a kind of sieves through which constituents of the blood could pass and give rise to different secretions. Against their view of the body as a sophisticated machine, Bordeu maintained that its functions depended on vital forces present only in the living organism. They could therefore not be reproduced outside the body.

Bordeu may seem somewhat vague and romantic in his concept of a vital force, but at the same time he had a remarkable ability to anticipate intuitively great discoveries that still lay in a distant future. For instance, he suggested that all major organs in the body function as glands by secreting substances specific for each organ to the blood, thus influencing each other by way of the circulating blood. He also proposed that it was the secretions of the gonads that were responsible for the sexual characters that develop in puberty. Here Bordeu in an ingenious way foresees what we now call endocrinology.

The ideas of vitalism had been received with enthusiasm in Germany where the concept of *Lebenskraft* ("vital force") was introduced by the brain anatomist Johann Reil. From the very beginning, German vitalism had a much more philosophic and mystical character than the French variety, and gradually it became the medical branch of natural philosophy, a school that flowered in Germany at the beginning of the 19th century. Its leading mind was the philosopher Friedrich Wilhelm Schelling (1775–1854), an outstanding pupil of Kant. At the early age of 22, the precocious Schelling published his book *Ideen zu einer Philosophie der Natur*, which became a canon of natural philosophy.

Théophile de Bordeu (1722–1776).
Courtesy of Bonnierförlagen, Stockholm, Sweden.

Schelling believed Nature to be posessed of a soul and in his opinion even inanimate material showed signs of life, as demonstrated by such phenomena as electricity and magnetism. His teachings aroused enormous, almost religious enthusiasm, not least among German physicians. Particularly appealing to doctors was of course the fact that Schelling considered medicine to be the foremost of all sciences and the one closest to the divine being. A central thought in natural philosophy was the polarity that characterized everything in the universe. Life was seen as oscillating between a positive and a negative pole, between the positive sun and the negative earth. The male character was mainly influenced by the sun, while the female nature was more earthly. Disease was considered to be caused by a disturbance of the natural polarity, but could alternatively be explained as a fall from a higher to a lower level in the hierarchic structure of creation, where man was at the top of the biological ladder.

A Memorial in Hotel-Dieu

Rummaging in odorous corpses to come up with new facts about diseases did not appeal to the fastidious minds that subscribed to the medical branch of German natural philosophy. Why bother with harsh realities when metaphysical speculations were much more pleasant and infinitely easier? Instead it was in France after the revolution that a young physician took over from the old Italian master and brought pathology one important step further.

Marie Francois Xavier Bichat (1771–1802) was born in the small village of Thoirette near the Swiss border. He was the son of a doctor but his early studies in Lyon included rhetoric and philosophy rather than medicine. However, at the age of 20 he took up medical studies and after a short period as assistant surgeon in the victorious armies of the republic, he became the favorite pupil of the famous surgeon Pierre Desault at the Hotel-Dieu in Paris. Desault recognized the exceptional qualities of his young assistant, but unfortunately their collaboration did not last very long. Already in 1795, Desault died unexpectedly after a short illness and Bichat was left with the sad duty of editing and publishing his beloved teacher's scientific manuscripts. His own life was destined to be short, but with his complete dedication to science and his tremendous capacity for work, he managed to make a lasting contribution to medicine in those few years.

In 1801, Bichat published his most important work, *Anatomie Générale*. Like his predecessor Morgagni, he emphasizes the leading role of the solid organs rather than the body fluids, both in the normal functions of the body and in disease. However, he goes one step beyond Morgagni and points to the tissues as the bearers of *les propriétés vitales* ("the properties of life"). Bichat realized that

Friedrich Wilhelm Schelling (1775–1854).
Courtesy of Bonnierförlagen, Stockholm, Sweden.

the organs were made up of many different tissues with distinct properties. One tissue in an organ might be affected by disease while the rest of the organ is relatively intact. For instance, he points out that in meningitis the membranes are the seat of the inflammation while the brain substance itself may be totally unaffected. Having thus created the modern concept of tissues, Bichat curiously enough contemptuously rejected the most important instrument for the closer examination of the tissues — the microscope. In his opinion it was completely worthless, and it must be admitted that the microscopes available at this time were not very advanced. It would be the following generation of pathologists that fully realised the capability of this instrument and thereby brought pathology to a new level of sophistication — cellular pathology.

The concept of vital properties is an essential element in Bichat's biological thinking, and in the more philosophical treatise *Reserches Physiologique Sur La Vie et La Mort* (*"Physiological Studies of Life and Death"*), he sums up his standpoint in the famous dictum: "Life is the sum of all functions that resist death." He also steadfastly insisted that *les propriétés vitales* were in principle different from the phenomena of the inanimate world and could therefore not be explained by the physical laws that applied there. Undoubtedly, there is something of vitalism in Bichat's general attitude and he praised the Montpellier medical school, that had been the stronghold of vitalism in France since the days of Théophile de Bordeu, for upholding its basic ideas. Maybe the fact that Bichat's father had received his medical training in Montpellier was significant here.

Bichat had a lasting influence on medical thinking in France and even Claude Bernard, certainly no admirer of the vague doctrines of metaphysical vitalism, paid tribute to the importance of Bichat's ideas. They became an inspiration also outside the boundaries of medicine, and the confirmed sceptic and pessimist Arthur Schopenhauer, insisted that he was a disciple of Bichat's. Even Napoleon would seem to have belonged to his admirers since when he heard of his death, he ordered a slab of marble with Bichat's name together with that of Desault to be put up in Hotel-Dieu. A remarkable tribute from the absolute ruler of France to a young physician, who at the time of his death held the lowly position of *médicin expectant* at that hospital.

Cellular Pathology

Because of the profound influence that Bichat had on leading French clinicians, pathological anatomy paved the way for new concepts and an entirely new attitude

Xavier Bichat (1771–1802).
Courtesy of Bonnierförlagen, Stockholm, Sweden.

in medicine. Instead of loose speculations built on fanciful generalizations and artificial medical systems, the new generation of clinicians tried to relate their observations at the sickbed to the actual findings at the autopsy. This was true not only in France but also in the German-speaking countries. At the famous hospital Allgemeines Krankenhaus in Vienna, Karl von Rokitansky took up the pathological anatomy that Morgagni and Bichat had introduced and even in Germany, where medicine had for so long been ensnared by the enticing doctrines of natural philosophy, the new ideas were gaining ground. The leading German clinician in the first half of the 19th century was Johann Lucas Schönlein (1793–1864) whose teaching was a curious mixture of old and new. He taught, for instance, that disease was caused by a lack of balance between the "individual" and the "planetary" principles in the organism — a wonderful example of the muddled thinking of natural philosophy when applied to medical problems. On the other hand, he introduced the new methods of physical diagnosis, percussion and auscultation in German medicine, and he was well aware of the importance of pathological anatomy for the clinician. It is significant that the greatest figure in modern pathology was one of his pupils.

A Young Revolutionary in 1848

Rudolf Virchow (1821–1902) was born in rural eastern Pomerania where his father had a small business. The economic circumstances of the family were modest but Virchow was able to attend the Gymnasium in Köslin where he showed proficiency both in the sciences and the humanities. He was considered so promising that like several other leading figures in German medicine, he was accepted for free medical studies at the Friedrich Wilhelm Institute in Berlin. Here he was trained in physiology by Johannes Müller and in pathological anatomy and clinical medicine by Johann Schönlein. He got his MD in 1843 and was appointed resident at the Charité Hospital in Berlin where he began his pathological research studying vascular thrombosis and embolism.

Virchow's superiors had obviously been impressed by his talents, since at the young age of 24 he was already invited to give a couple of lectures to a large and distinguished audience at the Friedrich Wilhelm Institute. It was not in Virchow's nature to be reticent and uncertain of himself, and he now stepped forward as the spokesman for a new and radical generation in German medicine. Unlike his mentor Schönlein, he completely rejected the metaphysical speculations of natural philosophy and instead emphasized that progress in medicine must be built on three main foundations: clinical observations with the aid of the new physical diagnostic methods; physiological and pharmacological experiments using

Rudolf Virchow (1821–1902).
Courtesy of Bonnierförlagen, Stockholm, Sweden.

laboratory animals; and pathological anatomy, where he stressed the enormous potential of the microscope. He also offended a great many venerable gentlemen in the audience by brazenly declaring that life was nothing but the sum of the physical and chemical processes in that basic living unit of the organism — the cell.

In 1847, he became an instructor under Johannes Müller at the Berlin University and was also appointed prosector (pathologist) at the Charité Hospital. To facilitate the publication of his radical ideas he started a scientific journal of his own, *Archiv für patologische Anatomie und Physiologie*, destined to become one of the leading medical journals of the century. In launching it, he declared in a kind of editorial manifesto that metaphysical speculations were out: "Now is not the time for systems, but for detailed investigations." He would soon prove to be as radical in politics as in his medical views.

When typhus fever appeared in Upper Silesia in 1848, he was sent there as a member of a governmental commission to investigate the outbreak. He was appalled by the social conditions he encountered, particularly among the Polish minority and his already liberal views were further strengthened. In his report to the Prussian government, Virchow strongly recommended not only medical measures but radical economic and social reforms. These ideas did not endear him to the authorities, who were shaken by the revolution that had just deposed King Louis Philippe of France and that now threatened to spread to other European countries. In March of 1848, the people of Berlin rose in revolt and it came to street fightings where Virchow appeared on the barricades together with the other revolutionaries. The uprising was soon put down by the army that was fiercely loyal to the king and generally a stronghold of the reactionary. In the end, Virchow paid the prize of his radical political activities and he was dismissed from his position at the Charité. He thought it better to leave Berlin and in 1849 he accepted a position as professor of pathological anatomy in Würzburg, the first of its kind in Germany.

The Triumph of the Microscope

To us, the cell is central for our understanding of the biological world. We entirely agree with the doctrine that there is no life outside the cell. However, this pre-eminence of the cell in biological thinking dates back no further than to the first half of the 19th century. Already Malpighi had seen cells in plant material, but this observation had largely been ignored until the English botanist Robert Brown found in the 1830s that plant cells contained a nucleus. At this time, the German botanist Matthias Schleiden had come to realize the fundamental importance of the cells and their nuclei for the life of the plant. He was a close friend of the anatomist Theodor Schwann, who as the result of a famous after dinner conversation with Schleiden was inspired to look for cells in animal tissues. In 1839, Schwann published the results of these studies that represent the real breakthrough of the cellular theory in biology. Nevertheless, as we have seen, he must share the credit with several other pioneers including the Czech physiologist and histologist, Johannes Evangelista Purkinje, who in 1837 described structures in animal tissues that he likened to the cells found in plants.

Unlike Bichat, Virchow clearly saw the importance of the microscope and its ability to bring pathological research to the level of the cell. It was a question of getting 300 times closer to the processes of life, as he used to put it. His enforced exile in Würzburg, that kept him away from politics, also gave him the time he needed to work out the basic concepts of his cellular pathology and initiate the publication of his *Textbook of Special Pathology and Therapy* in six mighty volumes. Eventually the Prussian authorities seem to have regretted their ill advised dismissal of Virchow, and in 1856 he returned in triumph to a chair in pathological anatomy at the University of Berlin. He soon gathered a group of outstanding young pathologists around him, including names like Julius Cohnheim, who investigated the microscopic characteristics of the inflammatory process, and Friedrich von Recklinghausen who first described the skin tumor that bears his name.

In 1858, Virchow published one of the most influential books in modern medicine, *Die Cellularpathologie in Ihrer Begründung auf Physiologische and Pathologische Gewebelehre* ("*Cellular Pathology as Based Upon Physiological and Pathological Histology*"). Here for the first time, cell biology (implying in those days the study of the morphology of the cell and its relation to cellular function in health and disease) had become the most important aspect of pathology. This meant an entirely new departure in medicine as a whole and the influence of Virchow on medical development for over a century was incomparable.

Schwann had believed that single cells originated from a kind of shapeless, unstructured cell-mass, but Virchow realized that all cells were the progeny of other cells — "*omnis cellula e cellula,*" to quote his famous dictum. He broadened this doctrine to include also the malignant cells. They arose from previously normal cells in the body; an idea that seems self-evident to us, but was quite original at the time. Modern cell biology is a direct continuation of Virchow's work, although with the important difference that we now try to explain cellular processes on a molecular level and are not confined to what we can see in the microscope as Virchow was.

Because of his deep feeling for the disadvantaged and underprivileged in society, Virchow had a tendency to emphasize social conditions as an important

Theodor Schwann (1810–1882).
Courtesy of Bonnierförlagen, Stockholm, Sweden.

factor in the outbreak of infectious diseases. His experience as a young man during the typhus fever epidemic in Upper Silesia no doubt played a role here. This may be one of the reasons why he tended to be sceptical of bacteriology and instead stressed the multifactorial nature of infectious diseases. Undoubtedly this attitude made him go too far in many cases. For instance, he stubbornly refused to accept the importance of Semmelweis's introduction of handwashing as a profylactic against puerperal fever. With all his admirable qualities, Virchow could sometimes be dogmatic and over-confident in his own infallible judgement and he had a tendency to become involved in acrimonious conflicts over both scientific and political issues.

A Cancelled Duel

His old political interest asserted itself, when in 1859 he was appointed to the Berlin City Council where he was active in promoting the improvement of the sanitary conditions of the rapidly growing city. In 1861, he was elected a member of the lower house of the Prussian Parliament as representative of a liberal party that he himself was a co-founder of. He became a leader of the opposition against Bismarck and disapproved strongly of the chancellor's military expenditures and his plans for the unification of all German states under Prussian leadership, that were to result in the creation of the German Empire after the victory against France in 1871. The conflict with Bismarck became so bitter that the enraged iron chancellor actually challenged Virchow to a duel. Fortunately for medicine, the duel was avoided through intervention on the highest level.

Virchow also found time for such interests as anthropology and archeology. He was a friend of the amateur archelogist Heinrich Schliemann and conducted digs both on the presumed site of the ancient Troy and in Egypt. On his 80th birthday in 1901, there were worldwide celebrations on a magnificent scale as a tribute to his internationally recognized standing as the renewer of pathology. During his long life, he had been instrumental in the victory of scientific medicine over the vague doctrines of natural philosophy that had captivated German medicine in his youth, and he had helped raise it to a dominating position in the world.

The New Physiology

It is not altogether easy to decide when physiology first emerged as a full-fledged science. Already Galen had made systematic physiological experiments, for

Claude Bernard (1813–1878).
Courtesy of Bonnierförlagen, Stockholm, Sweden.

instance in neurophysiology where he certainly was a pioneer. On the whole, it would seem that Harvey's discovery of the blood circulation is the best choice; when all the fantastic speculations that had littered this field for ages were suddenly swept away, and in their place a logical structure, based on sound experimental work was erected. However, all was not rational and built on a critical evaluation of experimental results and clinical observations in 17th century medicine. There was also a preference for easy generalizations that attempted to find a common denominator for everything, to provide an answer that covered all problems. The dominating school of thought at the time, the iatrophysicists or iatromechanics, had tried to reduce the living organism to a complicated machine that could be understood in purely mechanical terms. Against this simplified and unsophisticated model, the new physiology argued that the living organism could not be described using analogies pertaining to ordinary machines in everyday life.

Claude Bernard (1813–1878) was born in a hamlet outside the village of St. Julien in the department of Rhône, where his parents were vineyard workers. He was brought up in very modest circumstances and his studies at local religious schools were not very satisfactory, at least not from a science point of view. At the age of 19, he was apprenticed to an apothecary near Lyon. His experience of practical pharmaceutical work there gave him a lifelong distrust of clinical medicine and drugs. It must have been with mixed feelings that he nevertheless decided to take up medical studies in Paris after having first tried his luck as an author of vaudevilles and dramas.

There was nothing about the medical student Claude Bernard which in any way suggested that he would become the founder of modern physiology. On the contrary, he was regarded as not very promising and his fellow students were greatly surprised when later on they witnessed his rapid rise to fame. In 1839, he was accepted as an intern at the Hotel Dieu Hospital by the famous clinician and physiologist Francois Magendie. The professor soon noticed the great proficiency of his new intern at vivisection experiments. Bernard became his assistant and when Magendie died in 1855, he was succeeded by his former collaborator as professor of medicine at the Collège de France. Bernard's lectures at the Collège were mostly concerned with introducing the listeners to recent developments in physiology with special emphasis on his own research. "Showing science in the making rather than science already made," as he used to say. They draw an audience containing many distinguished scientists from all over the world and were published as a series of "Lecons" by the Collège. Bernard received numerous honors including being made commander of the Legion of Honour, senator of the Empire and a member of the Académie Francaise, whose president he also became.

In his doctoral thesis of 1843, Bernard showed that the saliva and the gastric juice contains enzymes that digest the food components. Thus, all carbohydrates are broken down to monosaccharides before they can be absorbed and utilized by the organism. A few years later, he was struck by the accidental observation in experiments on dogs and rabbits that the chyliferous vessels, which carry the absorbed fat from the intestines, were filled with fat below the point where the pancreatic duct opened into the intestine, but contained no fat above this point. He correctly concluded that the pancreatic juice digested and saponified the fat, and together with the bile emulsified it so that the fat could now be absorbed throw the intestinal wall and appear in the chyle. Furthermore, it is obvious that Bernard realized that what he called "nutrition" was an extremely complex process involving also intermediary metabolism, to use a modern term, with its intricate system of innumerable enzyme catalyzed chemical reactions.

Magendie had shown that the blood contains sugar and he had naturally concluded that the blood sugar was derived directly from dietary sugar. In 1848, Bernard found somewhat unexpectedly that the blood sugar level was constant and independent of food intake. Instead, it seemed that the liver produced sugar which appeared in the blood. Later, he performed experiments with isolated livers perfused with water. He found that the still warm liver under these conditions continued to produce sugar but that eventually, as the liver cooled down, the perfusing water became free of sugar. However, if the liver was warmed up again, it once more started to produce sugar. Bernard concluded that the liver must contain a special "glycogenic" substance, and the same year he reported the discovery of "animal starch" (glycogen) in the liver.

Another important finding was that, contrary to Lavoisier's belief that heat was produced exclusively in the lungs by a combustion process, this took place in all tissues. Bernard also realized that such "vital combustion" was not just a direct burning of organic material to produce heat and carbon dioxide, like in the burning of a piece of coal in an oven. Instead, it was a very complicated process in many steps that required the presence of special enzymes. With that, Bernard can be said to have grasped the fundamentals of energy metabolism. In fact, if one takes his whole immense life's work into consideration, he has undoubtedly contributed as much to biochemistry as to physiology.

Harvey's discovery of the blood circulation had established the vascular system as being made up of arteries and veins, later shown by Malpighi to be connected by capillaries. However, this was a static system in the sense that the vessels were not thought to be able to vary their diameter in order to regulate the blood flow to the tissues. When Bernard found that special nerves controlled the flow of blood by either constricting or dilating the vessels, it added a new dimension to our understanding of the circulatory system.

Hermann von Helmholtz (1821–1894).
Courtesy of Bonnierförlagen, Stockholm, Sweden.

The idea of a *"milieu intérieur"* is central in Bernard's thinking about the requirements for life. He considered these problems for many years although the term as such was not coined until 1857. Consequently, its connotation varied somewhat over the years and it is not altogether easy to give an unambiguous definition of the term. In any case, it involves the idea of a stable interior environment that nourishes and protects the living cell. According to Bernard, life can be seen as a permanent conflict between the cells that constitute the organism and the outer world. Without the *"milieu intérieur,"* life would not be possible. Obviously, what Bernard had in mind were complicated, multicellular organisms — the microorganisms represented an unknown world when he first conceived of his *"miliéu interieur."*

The Indestructable Energy

In the beginning of the 19th century, German physicians were so enthralled by the alluring doctrines of natural philosophy that it was difficult for scientific medicine to hold its own in the prevailing atmosphere of romantic daydreams. The man who must be credited with having upheld the rule of common sense in German medicine during this period of confusion and fanciful theories is the physiologist Johannes Müller (1801–1858). This is all the more remarkable, as in his younger days, Müller was a fervent believer in natural philosophy and throughout his life retained a vaguely vitalistic outlook. His greatest importance was as an inspiring teacher who formed a school of enthusiastic and outstanding disciples around him, that included such names as the great pathologist Rudolf Virchow and the eminent physiologist and physicist Hermann von Helmholtz (1821–1894).

Helmholtz came from a middle-class family in stringent economic circumstances. His father was a poorly paid teacher in Potsdam outside Berlin. Hermann was a sickly child who suffered from hydrocephalus as a result of meningitis and was bedridden for long periods of time. Like many other great scientists, he found the Latin swotting of his early schooldays a nuisance, but in the Gymnasium he was interested in physics and wanted to study the subject at the university. However, his father could not afford this on his poor teacher's salary and instead he persuaded his son to enter the Friedrich Wilhelm Institute that provided free medical studies for boys on the condition that they agreed to serve ten years as doctors in the Prussian army.

During his medical studies, Helmholtz began research in physiology under Johnnes Müller, where he became acquainted with other disciples of Müller's like Emil du Bois-Reymond, Ernst Brücke and Karl Ludwig, who would all become outstanding scientists that helped rejuvenate German physiology and free it from the fetters of vitalism and natural philosophy. When Helmholtz received his MD in 1842, he was appointed surgeon to a hussar regiment in Potsdam, a somewhat peculiar position for a man who was destined to become one of the greatest physiologists and physicists ever. He still maintained connections with the Müller school in Berlin, but even so, the hussar regiment was hardly an intellectually stimulating environment. However, Helmholtz was an introspective and thoughtful young man, perhaps as the result of his sickly and bedridden childhood, and he did not seem to have suffered unduly from his intellectual isolation. He had frequently mused on the problems of life and death and what distinguished these states from each other — not a very unusual thing for a young man in romantic Germany at the time. The difference was that, while in many of his contemporaries, such thoughts produced nothing but elegiac outbursts of Weltschmerz in sundry poetry albums, in Helmholtz they led to the ingenious concept of the conservation of energy.

The vitalists believed that the living organism was possessed of a mystical vital force that was ultimately responsible for all functions of life and that disappeared with death. Helmholtz rejected these ideas since he realized that the assumption of a vital force in reality meant that you accepted the possibility of a perpetuum mobile. Instead, he proposed that both body heat and mechanical work performed by the animal was derived from the chemical energy released by the combustion of food. In 1847, he published his seminal paper *On the Conservation of Energy*, but it was some time more before the scientific community fully realized the importance of his new thoughts. In fact, these thoughts of Helmholtz's were not all that new. An obscure German physician, Robert Mayer, had come to the same general conclusions some years earlier while he was a ship's doctor in East Indian waters. One must marvel at the creative thinking of these young German doctors, on a subject that appears not very close at hand either for a ship's doctor or a surgeon of hussars.

The next year Helmholtz succeeded his friend Ernst Brücke as professor of physiology at Königsberg where he would remain for six scientifically very productive years. He managed to measure the velocity of the nerve impulse, something that the vitalists held to be impossible since it was supposed to be a spiritual process. In 1851 followed the construction of the ophthalmoscope prompted by his pondering of the old observation that the eyes of animals can be seen to glow in the darkness when hit by light. Helmholtz realized that this must mean that light entering the eye through the pupil is reflected from the inner wall of the eye. With an appropriate optical instrument, it should therefore be possible to inspect the retina. In a little over a week's time, he had constructed the ophthalmoscope, an instrument that not only revolutionized ophthalmology but also meant that the condition of the patient's blood vessels could be evaluated, for instance in diabetes or hypertension.

In 1858, Helmholtz was called to Heidelberg as professor of physiology and he remained there until 1871 when he accepted the prestigious chair in physics at the University of Berlin. During the Heidelberg years, he made fundamental contributions to physiological acoustics and to the problem of color vision, treated in his monumental *Handbook of Physiological Optics*. In Berlin, he was active exclusively in physics.

Hermann Helmholtz and Claude Bernard represent two entirely different types of scientists; each of them had a fundamental influence on the development of physiology. Bernard is the great experimentalist whose pioneering work vividly illustrates the importance of animal experiments in modern physiology. Helmholtz, on the other hand, shows how close the problems of physiology can be to those of theoretical and experimental physics. Even as human beings, these great figures were strikingly different. Bernard was brilliant both as a lecturer and a writer, while Helmholtz was considered a dull lecturer and completely lacked the literary talent that Bernard had earlier demonstrated as a young man. Helmholtz himself has told of the endless trouble that he took over his publications, rewriting them over and over again, until he was finally satisfied with his work.

Bankruptcy and a Gleam of Hope

It is hardly an exaggeration to say that when scientifically based clinical medicine slowly began to emerge during the first decades of the 19th century, the practice of medicine was in a state of bankruptcy. The intellectual framework that had sustained medical thinking for over 2000 years, the legacy from Hippocrates and Galen, was in a state of chaos and disintegration. Nor was that all. Bloodletting, the firm and never questioned mainstay of all therapy, was tottering to its fall — the greatest therapeutic upheaval ever in the history of medicine. In fact, around the mid-century, this ancient treatment disappeared from the medical scene as through a trapdoor, and in its fall it dragged with it the whole venerable edifice of Hippocratic medicine. The problem was that its demise left a therapeutic vacuum so that physicians became helpless victims of clever and merciless caricaturists like Honré Daumier. Never before or after has the practicing of medicine been held in such low esteem as in the middle of the 19th century. However, there was a gleam of hope in all that darkness on the clinical scene in the form of a great prophylactic breakthrough at the end of the 18th century.

Variolation

Among epidemic diseases, smallpox have perhaps been the most dreaded. This scourge seems to have been endemic in the Far East from time immemorial, and with the Arabic conquest it spread to North Africa and from there to Spain. The Islamic physician Rhazes, who had earlier described the disease in the beginning of the 10th century, knew that the infection left the surviving patient with a lifelong immunity. He seems to have regarded it as a comparatively mild children's disease (he did not distinguish between smallpox and the measles) which almost everyone had to go through. In fact, he saw it as a physiological adjustment of the body fluids, a kind of maturation process, and this view of smallpox seems to have prevailed as far as to the renaissance.

On the other hand, the terrible ravages of the smallpox in the Americas during the 16th century, when large parts of the native population were wiped out, shows the disease from a horrifying side. From the 17th and 18th centuries, we have detailed reports of epidemics at regular intervals in Europe with mortalities ranging from 15 to 30% in most instances. However, in the most terrible form of the disease, variola haemorrhagica, with bleeding from the pustules, the mortality could reach close to 100%. It was not only the high mortality that made the smallpox so dreaded, it also left their surviving victims with disfiguring scars and often resulted in disability by blindness, that was a frequent complication because of the tendency of the skin eruptions to engage the cornea.

European physicians were completely helpless and could neither treat nor prevent the terrible disease, but in the Far East, unknown to western medicine, a prophylactic method, variolation, had been practiced for centuries. It had been found that if a child (smallpox was still primarily a children's disease) was inoculated in a scratch of the skin with pus taken from a fresh pustule of a smallpox case, the child got the disease in a mild form that did not leave any scars and had a low mortality compared to that of smallpox contracted in the usual way.

Lady Mary Wortley Montagu, famous English letter writer known for her wit and beauty, is often credited with having introduced variolation in Europe. In 1716, her husband became ambassador at Constantinople and his young wife accompanied him there. During her stay in the Ottoman capital, she must have heard of variolation and in 1717, she had her three-year-old son inoculated with smallpox. In her vivid and colorful letters home, she described the successful variolation and this caused considerable interest and commotion there. On her return to England, she also had her five-year-old daughter inoculated, which drew further attention to this newfangled method of smallpox prophylaxis. There were some protests on religious and ethical grounds, but the royal family became interested, and after having tested variolation on half a dozen criminals and some pauper children with satisfactory results, the Prince of Wales had his two daughters inoculated.

When this proved a success, variolation became very popular in England and was widely used throughout the century until Jenner introduced his vaccination. From England, inoculation spread to France and the rest of the European continent and it was also early on introduced into the English colonies in America. The method represented the first systematic attempt to a smallpox prophylaxis, even if its effectiveness has been questioned by some modern authorities. At the same time, it had some serious disadvantages. Variolation had a mortality that was not negligible, even if it was only a small fraction of that experienced in smallpox epidemics, and it also undoubtedly could lead to the spreading of the disease. It might have reduced the frequency of smallpox epidemics where it had been introduced, but it was far from the ideal method of prophylaxis.

A Country Doctor in Gloucestershire

Edward Jenner (1749–1823) was born in the little town of Berkeley in Gloucestershire, and to this rural idyll he returned after his medical education in London and set himself up as a medical practitioner. Here he would have remained, happy and content though anonymous, had it not been for a fascinating epidemiological observation known to the local farmers for generations.

The cows in Gloucestershire often suffered from an illness with characteristic eruptions on the teats. This relatively benign condition was called cowpox and could be transmitted to the hands of milkers where it caused similar skin eruptions as on the teats of the cows. It was a general belief among the local population that people who had been infected with cowpox were afterwards resistent to smallpox. In fact, unknown to Jenner, a farmer in the neighborhood prophylactically inoculated both his wife and their four children with cowpox. Jenner became deeply interested in the ability of the cowpox to protect against smallpox, but he found little interest and support among his medical collegues, who refused to believe in the old medicine folklore about cowpox.

Finally, in May of 1796, having pondered this problem for more than two decades, he was ready to make a daring experiment. Using liquid from a pustule on the hand of a milkmaid with cowpox, he inoculated an eight-year-old boy, James Phipps, in a superficial scratch on his upper arm. Six weeks later, Jenner tried to inoculate the boy with smallpox, but James proved to be resistant to variolation, in the same way that Jenner had found to be the case with people that had previously had cowpox. A year later, he sent a report of his experiences with

Edward Jenner (1749–1823).
Courtesy of Bonnierförlagen, Stockholm, Sweden.

cowpox inoculations as a prophylaxis against smallpox to the Royal Society in London, but they refused to consider it. The Society had previously accepted a report by Jenner on the breeding habits of the cuckoo, but these fantastic ideas about cowpox seemed altogether too incredible.

In 1798, Jenner published a small book with the title *An Inquiry into the Causes and Effects of the Variolae Vaccinae* where he described 23 cases of inoculations or spontaneous infections with cowpox, which he claimed to have resulted in immunity against smallpox, as indicated by resistance to variolation. Initially, there was a certain reluctance to accept this new smallpox prophylaxis, that became known as vaccination after the latin word for cow, *vacca*. One reason may have been the unfortunate experience of a colleague who had his vaccine contaminated with smallpox virus. This caused at least one death among his patients, but more important was probably the general feeling amongst the public that it was degrading and ridiculous for human beings to be inoculated with material from cows. Talented caricaturists surpassed each other in depicting the laughable and mind-boggling complications that might result from Jenner's vaccination, where the poor victims could become more or less cow-like.

However, this initial resistance was short-lived and the vaccination against smallpox spread rapidly all over the world. Soon, the dreaded disease had almost completely disappeared in Europe. Parliament on two occasions voted Jenner large sums of money, Napoleon had a medal struck in his honor in 1804 and later, on Jenner's request, freed several British citizens who had been interned in France during the war. Jenner had had the courage to take up an observation in folk medicine and develop it into a prophylactic method of immense importance. He clearly recognized the enormous potential of his vaccination and correctly predicted that it would one day completely eradicate smallpox from the face of the earth. On the other hand, there is nothing to indicate that he had a deeper understanding of the immunological principles underlying his method than the Gloucestershire farmers who had made the original observation, but in terms of practical results and lives saved there is no one who can compare with Jenner.

A WORLD UNKNOWN

Contagion Versus Miasma

In ancient medicine, disturbances of the four body fluids were seen as a sufficient explanation of all diseases, and this view hardly encouraged the Hippocratic physicians to study infectious diseases as distinct entities in the medical panorama. In spite of their otherwise complete faith in Hippocrates and Galen, the medieval physicians were the first to recognize that certain illnesses, particularly those that took the form of periodically occurring, devastating epidemics, were caused by what we call infections, although they used the term contagion. It was also during this otherwise rather undistinguished period in medicine that the first, albeit not very effective steps were taken to contain epidemics by quarantine and isolation of the sick.

Seeds of Infection

The nature of the agent that supposedly transmitted sickness from one individual to another, the so-called contagium, remained an enigma until what seemed to be a new disease, syphilis, made its appearance in Europe in the 16th century. It was believed to have been brought to Europe from the Americas by the returning Spanish conquistadors and was therefore often called the Spanish disease. On the other hand, it was also called the French disease (*morbus gallicus*) which illustrates the tendency to blame hostile nations for the terrible illness. The only remedy seemed to be preparations containing mercury, but often enough the cure was worse then the sickness itself.

The disastrous outbreak of syphilis had consequences also for the general attitude and social life of renaissance man. In the Middle Ages, people were used to an existence in closely knit communities marked by very intimate contacts between individuals. Plague and other epidemics certainly had a disruptive effect on the medieval society while they lasted, but syphilis was something different. It was a disease that soon became endemic, i.e. it established itselt permanently all over Europe, and it was linked to the most intimate of all human relations. One can understand if it changed the whole cultural and social pattern of the old society. Characteristically enough, it did not really make people less promiscuous, but it probably was a decisive factor in putting an end to such activities as the communal baths, something that hardly improved the personal hygiene of the period.

Nowadays, it is unusual to find a great poet among the pioneers of biomedical research, but during the renaissance things were different. The Italian poet and physician Girolamo Fracastoro (1478–1553) owed a considerable part of his

Girolamo Fracastoro (1478–1553).
Courtesy of Bonnierförlagen, Stockholm, Sweden.

reputation as a master of Latin verse to his epic poem *Syphilis sive Morbus Gallicus*. The shepherd Sifilo had offended Apollo and the revengeful god punished him with the dreadful disease that takes its name after the poor victim of Apollo's wrath. In his poem, Fracastoro described the symptoms of syphilis eloquently and in considerable detail that testifies to his familiarity with it. However, his interest in infectious diseases was not limited to syphilis. He had studied tuberculosis thoroughly and had noted that it could be transmitted not only by direct contact with the patient but also through bedclothes and garments previously used by the sick. These observations led him to suggest in his famous book *De Contagione et Contagiosis Morbis*, printed in 1546, that contagious diseases were caused by invisible particles, *seminaria contagiosa* ("seeds of infection"), that he considered to be self-replicating and specific for each disease. He also pointed out that these *seminaria* differed in their ability to survive outside the body of the patient, those causing tuberculosis being particularly difficult to get rid of.

The fame of Girolamo Fracastoro in his own time was as a poet and not as a scientist and physician. His theory about *seminaria contagiosa* was far ahead of his time and never had any real influence on contemporary medicine. Recognition of his greatness as the first physician to have an inkling of the microbiological world would not come until the triumphal arrival on the medical scene of this new science in the second half of the 19th century. Today, we realize that he is one of the great sages in the history of medicine, remarkable because, unlike so many other medical theoreticians before and after him, he built his theories on thorough, critically evaluated clinical observations — and because in the end he turned out to have been right.

The idea of a contagium is thus a very old one, but even in the first half of the 19th century its proponents, the contagionists, had very conflicting notions of its real nature. Some held that it was a chemical compound with different composition for different diseases. Another variation of this chemical theme was that the contagium was really an abnormal product of the metabolism of the diseased organism. The other main theory was of course that of Fracastoro with replicating but invisible particles as the vehicles of infection. In 1840, the German anatomist Jakob Henle (1809–1885) published his remarkable *Patologische Untersuchungen* ("*Studies in Pathology*") where he reasoned, completely without any experimental support, that living organisms with the ability to replicate were the cause of contagious diseases. His argument was that only such organisms could explain how a minimal amount of contagious material can multiply in the infected body and cause disease. Thus, Henle can be said to have provided the theoretical foundation for the development that would follow in the second half of the 19th century.

A Poisonous Effluvium

The anti-contagionists took a very different view of this whole problem. They followed in the footsteps of Hippocrates by stressing the importance of the environment. The quality of the air and the water and of the living conditions in general influenced the state of health, but the anti-contagionists went further than that. Something emanating from the earth itself but possibly also influenced by the celestial bodies, an effluvium they termed miasma, pervaded the air and caused disease. The nature of the illness was ultimately determined by the character of the miasma that prevailed in a certain region.

They were fascinated by what they called *constitutio morborum* or the "constitution of the disease," which could change with time so that a particular disease might have a certain character one year and another a couple of years later. At one time, the spectrum of diseases was dominated by the phlegm, at another it was mainly bilious, etc. This is of course reminiscent of the four body fluids, but the ruling principle was that of miasma that supposedly caused these changes in the constitution of diseases.

The concept of miasma led to endless speculations about its true nature and what might influence its character. Apart from the properties of the geographic location itself, celestial phenomena like comets or eclipses of the sun or the moon were seriously believed to influence the miasma, but the main factor was perhaps the weather. In the first half of the 19th century, official reports about outbreaks of epidemics almost invariably began with lengthy considerations of the prevailing meteorological conditions and their role in the catastrophe. The four seasons and their influence on the frequency of contagious diseases was also emphasized by the advocates of miasma, and here they were on somewhat firmer ground. It was, after all, well established that respiratory diseases were more common in the winter, while gastroenteric infections were favored by the summer heat, etc.

With the benefit of hindsight, we could of course marvel at the success of the miasma theory. All the facts should have argued in favor of the contagionists, one might have thought. However, the problem from their point of view was that serious and widespread diseases showed properties that apparently were much more consistent with the miasma hypothesis than with the idea of living seeds of infection which were transmitted through personal contact with the sick. The classic example of this is of course malaria, where the name of the disease means "bad air," in itself an indication of a miasmatic origin. The spreading of the disease through something in the air (the mosquito carrying the malaria plasmodium in its salivary glands), that is present mainly in certain geographic regions (those that favor the reproduction of the mosquito), certainly agreed in every respect with the teachings of the anti-contagionists. There was, however, another epidemic disease rife in Europe, the study of which would settle the dispute in a somewhat unexpected way.

Cholera and the Final Verdict

Malaria had been known in Europe since time immemorial, while cholera until the beginning of the 19th century had been confined to Asia, where it seemed to be endemic in certain parts of India, in particular the Bengal region between the rivers Brahmaputra and Ganges. From here it periodically spread to other Asian countries, but it had never entered Europe until in 1830 it suddenly broke out in the Russian town of Nizhni-Novgorod, at the great market that was held there. It then overrun all of Europe and even took ship to the United States, causing distress and panic wherever it appeared. The mortality was very high and on the continent of Europe, where the contagionists had the upper hand, the authorities took the most brutal measures with quarantines and strict isolation of the sick, to prevent the disease from spreading to previously uninfected areas, but to no avail.

On the other hand, in the British Isles, where miasmatic opinions prevailed, quarantines and isolation of infected districts were not employed, since it was considered both inhuman and bad for trade. Here the anti-contagionists concentrated on such measures as building new hospitals for the cholera victims and improving the general hygienic standard, including the supply of good quality water. Nevertheless, the ravaging of the cholera was no worse in Britain than on the continent. Furthermore, the ability of the cholera to infect previously healthy people, who had had no direct contact with the sick and the fact that it tended to be restricted to certain limited areas, such as a particular quarter in a big city, seemed to be more consistent with a miasmatic explanation of the disease. Eventually, the opinion on the European continent and particularly in France, swung from contagionism to a fervent belief in miasma.

Paradoxically it was a physician in anti-contagionist England who vindicated contagionism by his studies of the epidemiology of cholera, a disease that seemed at the time firmly established as being of miasmatic origin. During an outbreak of cholera in London in 1849, John Snow (1813–1858) made a crucial observation that definitely turned the table on the anti-contagionists. As usual, the cholera was confined to certain districts and Snow found that in one of those, only people who fetched their water from a particular well became ill. When he looked into the matter, he found that the well was contaminated by the ooze from a nearby latrine. This had been going on for quite some time without any disastrous

Alphonse Laveran (1845–1922).
Courtesy of the Nobel Foundation, Stockholm, Sweden.

results, but Snow discovered that a few days before the local outbreak, the latrine had been used by an infected visitor from a distant part of London, where cholera was rife. The conclusion was obvious. The stools of the cholera victims contained the contagium, which because of poor hygiene was disseminated in the water supply and infected the local population which drew water from an infected well. In a way, both the contagionists and the miasmatics could be said to have each held part of the solution to the problem. Cholera was undoubtedly caused by a contagium, but better general hygiene and a good water supply was indeed crucial to combat the disease. However, on balance the contagionists must be said to have been victorious and the heyday of the miasma was over.

Miasma Revisited

To the anti-contagionists, malaria had been a prime example of a disease with a miasmatic origin. It would take a long time before the true nature of the malaria miasma was finally realized. In 1880, the French physician Alphonse Laveran (1845–1922), working as an army medical officer in Algeria, could demonstrate the presence of a plasmodium, later found to be the cause of malaria, in the blood of patients with the disease. Laveran's superiors do not seem to have been very impressed by his discovery and he was transferred to France, which effectively put a stop to any further study of malaria. It was not until he left the army and in 1896 obtained a position at the Pasteur Institute that his contributions were really appreciated. In 1907, he was awarded the Nobel Prize for his work on pathogenic protozoa.

In 1897, the British military physician Ronald Ross (1857–1932) found the plasmodium in the stomach of Anopheles mosquitoes that had sucked blood from malaria patients. He then demonstrated that the avian form of the disease could be transmitted through the bite of mosquitoes. The possibility of the mosquito being a malaria vector had been suggested previously, for instance by the English physician Patrick Manson, but Ross was the first to have definitely proved this hypothesis. He also showed that the most effective measure in order to eradicate malaria was to destroy the larvae of the mosquito, thereby preventing its replication. In 1902, he was awarded the second Nobel prize in physiology or medicine for his work on malaria.

This was truly the heyday of military medical officers. Never before had they played such an important role in research as now. When the American army occupied Cuba in 1900 after the Spanish-American war, the old scourge of Central America, the yellow fever, claimed numerous victims among the soldiers. A commission led by Major Walter Reed (1851–1902), and including several young

Ronald Ross (1857–1932).
Courtesy of the Nobel Foundation, Stockholm, Sweden.

bacteriologists from the Johns Hopkins Medical School, was sent to Cuba to tackle the problem. A Cuban physician, Carlos Finlay (1833–1915), had proposed already in 1881 that the mosquito *Aëdes aegypti* was the vector that carried the agent of the yellow fever. The commission now started a series of heroic experiments where its members together with volunteering soldiers exposed themselves to all kinds of hazards that might lead to infection with yellow fever. Some slept in the beds of deceased patients without any ill effects, but others that allowed themselves to be bitten by the suspected mosquito became seriously ill and one member of the commission, Jesse Lazear, actually died. In 1901, the commission could report that the mosquito was indeed the vector and that yellow fever could only be transmitted through mosquito bites, not by personal contact with the sick. The cause of the disease could pass through a bacterial filter, i.e. it was what we today call a virus. The hygienic measures taken, attempts to eradicate the mosquitoes and the use of nets to protect humans against them, proved very effective and yellow fever was brought under control.

The Birth of Microbiology

The leading scientific academies in Europe had long been in the habit of trying to stimulate research by announcing prize competitions, where they demanded answers to certain topical questions. The French Academy of Science was no exception to the rule, and in 1859 it announced a competition that concerned the question of whether spontaneous generation of life was possible. The problem of the generation of live organisms from inanimate material had been pondered by scientists for centuries and the most fantastic ideas had been entertained. According to the more conservative views, small organisms (including what we call microorganisms) could arise spontaneously in rich nutritional substrates like bouillon and blood, if these were kept long enough under favorable conditions. This idea seemed to be supported by the old observation that such substrates, even if they were originally clear, would eventually become cloudy, and on inspection in the microscope living organisms could be demonstrated. On the other hand, the Italian priest and biologist Lazaro Spallanzani (1729–1799) had shown already a hundred years earlier that if the vessel with the substrate was thoroughly boiled and then sealed off from the air, no living organisms would be formed. Nevertheless, the old question that would not go away came up again, and this time it so happened that among the people that responded to the invitation of the academy was one of the greatest medical scientists ever, even if he had never had a single day of formal medical education.

Louis Pasteur (1822–1895).
Courtesy of the Royal Swedish Academy of Sciences,
Stockholm, Sweden.

Louis Pasteur

Pasteur was born in the little village of Dole in the South of France in 1822 as the son of a tanner and former sergeant. His father had been one of Napoleon's veterans, the *vieux grognards*, and had been decorated on the battlefield for valour with the Legion of Honour. Perhaps this was of some importance for the French patriotism that was such a characteristic trait in his famous son. Pasteur was not a particularly successful pupil in the Latin swotting school of those days. Like many other great men, he seems to have been a late bloomer. When he got his baccalauréat from the Royal College of Becancon in 1842, it was with a "mediocre" in chemistry. A truly remarkable character which must have caused his teacher to have second thoughts, if he lived long enough to follow the career of his former student.

Pasteur then studied in Paris where he obtained his degree as doctor of science, and already in 1848 he made his first important discovery while studying the optical properties of tartaric acid. He found that the racemic, optically inactive form of the acid gave equal amounts of left-handed and right-handed crystals, distinguishable from each other. By isolating individual crystals of the two forms, he could show that they had the same degree of optical activity but with opposite rotation so that the activities cancelled each other in the racemic form. This discovery made him famous. He was rewarded with the coveted red ribbon of the Legion of Honour, and in 1854 he became professor of chemistry at the University of Lille.

Before he responded to the prize question of the academy, Pasteur had previously worked with problems in microbiology. He had become interested in the question of why alcohol fermentation frequently and inexplicably went wrong and produced sour beer or wine. Pasteur demonstrated convincingly that a successful fermentation required the presence of normally growing yeast cells and that failure was caused by infection of the brew with other microorganisms. This led him to believe that diseases in animals and humans might also be caused by infections with microorganisms. The idea was far from new or original, but unlike its previous proponents Pasteur would provide solid experimental proof of his theory.

In the meantime, Pasteur had moved to Paris where in 1857 he became director of scientific studies at the École Normale. His previous work on fermentation made him well qualified to take up the old question of spontaneous generation of life posed by the academy, even if he had no formal schooling in biology or medicine. Beginning in 1860, Pasteur now performed the same kind of experiments as Spallanzani, only many more and done under the most varying conditions. His capacity for work was enormous; even as an old man and partially paralyzed by a stroke which he had suffered at the early age of 46, he habitually ended his conversations with his collaborators by repeating: "Il faut travailler, monsieur!" ("One must work, Sir!"). His results definitely showed that living organisms could not arise spontaneously in sterile substrates, provided that the vessels were sealed off from the air. This settled the question and confirmed Spallanzani's conclusions, even if it took some time before the impossibility of a spontaneous generation of life was generally accepted. All previous results to the contrary had simply been caused by infections with microorganisms from the outside.

In the 1860s, a disastrous sickness among the silkworms in the south of France was ruining that important industry and Pasteur was finally persuaded to investigate it, although to begin with he had declined with the argument that he had never seen a silkworm before. The results further increased his fame and in 1867 he became a professor of chemistry at the Sorbonne, the highest position that a French chemist could aspire to. In the following decades, his research became more and more oriented towards the prevention of illnesses in domestic animals caused by infections with microorganisms.

The professors at the medical schools tended to look down on this upstart of a natural scientist, who had no medical education whatsoever. They simply could not imagine that such a person might contribute anything of value to their own, jealously guarded field. It would be a long time before conservative physicians fully realized the revolutionary effect on medicine of Pasteur's venture into the unknown world of microbiology. Their attitude was a contributing reason why Pasteur for a long time concentrated his efforts on veterinary medicine, where the outlook obviously was less parochial. However, he was also motivated by his strong patriotism and the wish to improve the economic situation in France after the disastrous war with Prussia in 1870/71 and the crushing indemnity imposed on the country by the peace treaty.

Essays in Immunology

The research that Pasteur now embarked on presupposed a technique for isolating different microorganisms in pure culture. The method he devised had the advantage of being simple, but at the same time it was somewhat unpredictable. It consisted of making serial dilutions of mixed cultures grown in liquid media, until a certain volume of the diluted culture contained on an average only one cell. From this single cell, a pure culture might then be obtained and the properties of a particular bacterium could be studied at will, including the effect of different methods of growth on its ability to cause disease. Hopefully, bacteria that had been attenuated by manipulating their growth conditions, might then be used to vaccinate animals against the sickness caused by the virulent form of the bacterium. Unlike his predecessor, Jenner, who had explored a chance observation and was only interested in the practical aspects of vaccination, Pasteur worked systematically on the basis of a carefully considered hypothesis and had a real interest in the theoretical aspect of the problem.

Anthrax was a serious and economically important illness in domestic animals like sheep and cows, and it could also attack human beings. Pasteur had a strong ambition to focus his research on problems of practical importance and here was a disease that was a scourge to agriculture in his beloved France. In 1877, he made anthrax his main project. He was not the first to have studied the anthrax bacterium, but when he tried to grow it under varying conditions, Pasteur made an important observation. The bacterium was very sensitive to growth above body temperature and at 42°C he obtained an attenuated strain that caused only a mild disease when animals were inoculated with it. Using his attenuated anthrax bacteria, Pasteur made a full scale vaccination experiment on a number of sheep, goats and cows in the little town of Melun where the local agricultural society had placed them at his disposal. In May 1881, he vaccinated half of the animals and a month later he inoculated the whole herd with highly virulent anthrax bacteria. The outcome was indeed sensational. The vaccinated animals remained healthy, with the exception of one sheep which probably succumbed to an unrelated illness, but all the unvaccinated animals either died or became seriously ill.

Encouraged by his success with anthrax and two other animal diseases, fowl cholera and swine erysipelas, Pasteur now turned to rabies, a dreaded illness that could be transmitted also to humans through the bites of afflicted dogs. Rabies had a long incubation time varying from several weeks to months and this made Pasteur believe that it might be possible to prevent the outbreak by vaccinating the patient in the free interval. He could demonstrate the presence of the rabies "poison" (in reality a virus) in the saliva of sick dogs, but from the neurological symptoms of the disease Pasteur reasoned that its agent must be present in the central nervous system. With brain tissue from rabid dogs, he could induce rabies both in other dogs and in rabbits. He then dried spinal cords from sick rabbits for varying periods of time, and used these preparations to immunize healthy dogs. For each inoculation, Pasteur used cord tissue that had been dried for shorter and shorter periods; the last being with highly infectious tissue dried only for a day or two. Nevertheless, the dogs did not become ill and they now proved to be immune against inoculations with brain tissue from rabid dogs.

Originally, Pasteur had intended to eradicate the disease by vaccinating all the dogs in France, an ambitious project indeed, which could not possibly have succeeded. However, in July 1885 a distraught mother came to him with her nine-year-old boy, who had been badly bitten by a rabid dog two days earlier. Having first consulted two physicians, who declared that nothing could be done for the boy, Pasteur immunized him using the same procedure that he had previously worked out on dogs. This must have been an extremely trying period, not only for the little boy, Joseph Meister, and his mother, but also for Pasteur himself. However, the boy remained healthy and became one of those famous cases whose names will never be forgotten by all who take an interest in the history of medicine. Pasteur's success with the prevention of rabies greatly added to his fame both at home and abroad. Donations of money came from all over the world and made it possible to establish an institute in 1888 that was named Institut Pasteur in his honor. He served as its director until his death, and it has remained a leading center for microbiology and immunology to this day.

Fanciful Theories

Pasteur was deeply interested in the theoretical aspects of immunity and it may be of interest to consider his ideas on the subject against the background of previous speculations. The explanation of the Islamic physician Rhazes for the lifelong immunity after smallpox, was entirely based on his Hippocratic thinking about the cause of the disease. He saw it as a natural maturation process to rid the young organism of an excess of moisture in the blood. When this adjustment had taken place, the child was of course immune against any relapses since the necessary maturing of the blood had already occurred. The greatest name in Islamic medicine, Avicenna, held similar views and believed that smallpox was caused by the menstrual blood that accumulated in the mother during pregnancy and was transmitted to the fetus. The magic and dangerous properties of menstrual blood is a superstitious belief since time immemorial and even the otherwise far-sighted Girolamo Fracastoro thought that the seeds of infection that caused smallpox had a special affinity for the residual menstrual blood in the young child. Consequently, after the infection there was a corruption of the remaining menstrual blood which would eventually appear on the skin as the characteristic pustules. As a result, the child was of course immune against smallpox for the rest of its life.

This idea, that resistance to an infectious disease is caused by the infection having consumed something in the organism that is necessary in order to sustain the disease, we will encounter also in Pasteur's thinking about immunity. Having had vast experience with growing microorganisms under different conditions, Pasteur had noted that after a rapid initial growth (the logarithmic phase), the increase in number of cells gradually levelled off and the culture went into a stationary phase. He believed that this was because the microorganisms had consumed certain specific factors in the medium necessary for growth. On the face of it a very reasonable hypothesis, but he went one step further and assumed that the same was also true in an infected animal. When the enigmatic growth factors had been consumed by the infecting microorganism, the animal was immune to re-infection with that particular microorganism. This was the reason why his vaccines of live but attenuated bacteria could give immunity. They did this by consuming the specific growth factors without causing disease.

This attractively simple hypothesis had to be abandoned when it was found that dead bacteria could also be used as effective vaccines. In fact, immunity would eventually prove to be one of the most complicated phenomena ever tackled in medical research.

Pasteur may have indulged himself in some wishful thinking here, but no one can deny that he is the father of both microbiology and immunology and arguably the greatest figure in biomedical research ever. The pioneer who first implicated the microorganisms as the cause of infectious diseases and could sustain this idea experimentally. When he died in 1895, after a life spent at the working desk, his scientific reputation was incomparable. On his 70th birthday, Lord Lister had said in his eulogy that medicine owed more to Pasteur than to anyone else and it is easy to agree with the great surgeon.

Bacteriology and the Nemesis of Surgery

Since time immemorial, surgery had been haunted by a terrifying spectre that seemed inseparable from all operations — surgical fever. With the introduction of general anesthesia in the middle of the 19th century, the number of major operations performed increased considerably, but the dreadful mortality due to surgical fever was as bad as before. The very fact that serious injuries (for instance compound fractures where bone fragments had penetrated the skin) were now routinely hospitalized rather than kept at home, considerably increased the risk of an often fatal infection spreading from one patient to another in the large, overcrowded wards. Infections of surgical wounds were such matter of course complications that doctors had come up with the theory of the good pus, pus laudabile, and believed it to be a necessary way by which the organism rid itself of poisonous humors. The laudible pus was in contrast to the stinking fluid that oozed from a gangrenous wound and inevitably heralded the death of the patient, unless he could be saved by an amputation.

Even under favorable conditions, during peacetime and in well equipped and staffed hospitals, the mortality of big amputations could approach 50%, and in times of war it was considerably higher. During the war between France and Prussia in 1870/71, the French military surgeons performed 13,373 amputations, and in 10,006 of these the patient died, i.e. the mortality was 75%. During the siege of Paris, it reached almost 100% in certain hospitals. The rigorous cleanliness and sterility at all operations, that we take for granted, was very far from being a matter of course in the operation theaters in the middle of the 19th century. The surgeon operated in an old bloodstained jacket, instruments were not properly cleaned — what would be the point of that, they would soon be as dirty as before — and the doctor might wash his hands *after* the operation, but certainly not *before*.

The gruesome spectre of surgical fever had an equally feared counterpart that haunted the obstetric wards — puerperal fever. It was in the struggle against this destructive angel that the first decisive steps were taken towards the introduction of aseptic techniques that attempted to keep both instruments and the hands of the surgeon, as far as possible, free of infectious microorganisms. This was before the medical importance of these organisms had been realized, and one reasoned in terms of "poison" rather than bacteria. Nevertheless, by the 18th century, successful attempts had been made in England and Ireland to reduce the incidence of puerperal fever simply by practising general cleanliness. Unfortunately, the inertia of the medical community had prevented this important initiative from spreading.

In America, the author and physician Oliver Wendell Holmes had, by purely theoretical considerations, come to the conclusion that post-partum infections were largely caused by the obstetrician carrying the "poison" from one patient to another when he made a series of vaginal examinations without washing his hands in between. Consequently, Holmes suggested that the physician should thoroughly wash his hands and soak them in chlorinated lime whenever he had been in contact with an infected case. The use of chlorinated lime had recently been introduced to prevent "cadaveric poisoning," i.e. infection of wounds incurred during autopsies. In 1843, he wrote an essay with the title "The contagiousness of puerperal fever" which he read to the Boston Society for Medical Improvement, where it was duly ridiculed by his colleagues. These perceptive, not to say prophetic thoughts of Oliver Holmes, that had so amused his Bostonian audience, would be completely vindicated by a careful clinical investigation performed by a Hungarian physician working in one of the obstetrical clinics of the Allgemeine Krankenhaus in Vienna.

The Disregarded Pioneer

Ignaz Philipp Semmelweis (1818–1865) was born in Buda, the old part of Budapest on the right bank of the Danube, the son of a prosperous shopkeeper of German origin. He went to Vienna in 1837, where his father wanted him to study law, but Semmelweis was more interested in medicine and instead entered the medical school. During his studies, he came in close contact with some of the leading medical figures in Vienna, Josef Skoda in internal medicine, the pathologist Karl von Rokitansky, and the dermatologist Ferdinand von Hebra, who all became his friends and supporters. He completed his medical studies in 1844 and after graduation worked for over a year with Skoda, who taught him diagnostics and statistics. In 1846, he became an assistant physician at the First Obstetrical Clinic, where the teaching of the medical students took place, while the Second Clinic was entrusted with the instruction of the midwives.

At this time, the mortality in puerperal fever at obstetrical clinics all over the world was appalling, and this was true also in Vienna. The surprising thing was that in the First Clinic the mortality was 13%, a fairly normal figure internationally, while the Second Clinic had a mortality of only 2%. The compassionate Semmelweiss was very much perturbed by the sound of the little silver bell that announced the arrival of the priest coming to give the extreme unction to the dying women. He racked his brains over the incomprehensible mystery of why this should happen so frequently in his own clinic compared to the other one.

Among his colleagues, including the head of the clinic, Johann Klein, a narrow-minded and ignorant reactionary and no admirer of Semmelweis, miasmatic ideas prevailed and differences in the "morbid constitution" at the two clinics was seen as the explanation. However, Semmelweis thought otherwise. He had recently lost a dear friend, the young pathologist Jakob Kolletschka, who had become a victim of "cadaveric poisoning" after a cut received during an autopsy of a patient who had died of puerperal fever. Semmelweis was present at the autopsy of his friend and having been thoroughly trained in pathology by von Rokitansky, he observed that the findings in Kolletschka's body were the same as those seen in cases of puerperal fever.

It dawned on him that the cadaveric poisoning, that had killed his friend, and the puerperal fever in the wards must have the same cause. He also realized that the crucial difference between the First and the Second Clinic, was the fact that the First Clinic taught medical students, who carried the poison of puerperal

Ignaz Philipp Semmelweis (1818–1865).
Courtesy of Bonnierförlagen, Stockholm, Sweden.

fever on their unwashed hands from the bodies in the autopsy room to the women in the wards. The pupils in the school of midwifery at the Second Clinic, on the other hand, did not participate in autopsies. In May 1847, he instituted a strict routine where the students were required to wash their hands thoroughly and soak them in a solution of chlorinated lime before each vaginal examination. The medical students naturally rebelled against this infringement of the academic freedom, but Semmelweis stood his ground and the new rules were implemented. As a result, in a month's time the mortality in the First Clinic fell to the same level as in the Second Clinic, and further efforts, including thorough cleaning of the instruments, brought it down to an even lower level, so that in the end puerperal fever was virtually eliminated in the wards where these methods were used.

One might have thought that Semmelweis should have been eager to publish these sensational results, but for some reason he held back. Perhaps he had a premonition of how they would be received. In the meantime, his friend von Hebra wrote two papers on his behalf explaining the cause and prevention of puerperal fever, and Josef Skoda pressed for an official commission to look into Semmelweis's findings. But this was just after the crushing defeat of the liberal revolution in Vienna in 1848 and the repression of the Hungarian uprising. The reactionaries were now firmly in power both in the imperial government and at the university, and they were not inclined to do an unknown Hungarian physician, perhaps even a dangerous liberal, any favors and nothing came of Skoda's efforts. The hostility of Semmelweis's boss, Johann Klein, did not help things either. The real blow came in 1849 when Klein refused to reappoint Semmelweis and instead had him transferred to an unsalaried position as instructor of midwives.

Eventually, Semmelweis yielded to the persuasions of his friends and agreed to present his results at a meeting of the Association of Physicians in Vienna, that took place in May 1850 and was presided over by his patron Karl von Rokitansky. As might have been expected, it was received with something less than enthusiasm and Johann Klein continued to try to obstruct Semmelweis's career. He had difficulties supporting his family and finally Semmelweis had had enough. He suddenly left Vienna without even telling his closest friends, and returned to Budapest where he was offered a position as chief of the maternity ward of the hospital in Pest. Here, he introduced his aseptic technique with great success, and in 1855 he was appointed professor of theoretical and practical midwifery at the university.

His lectures attracted great audiences of interested students, but it was not until 1861 that he published his famous book, *The Etiology, Definition and Prevention of Puerperal Fever*. The reviews of the book in international journals were in many cases far from favorable and Semmelweis was drawn into a number of acrimonious scientific conflicts that made him grow bitter against the whole medical community. He came to see himself as a messianic figure, whose mission was to save the poor women in the maternity wards from the ignorance and callousness of intractable obstetricians. Semmelweis became increasingly unbalanced, and in 1865 he showed signs of being mentally ill. His friends persuaded him to come to Vienna, and when his condition deteriorated von Hebra had him confined to a mental hospital. In the sorrow and confusion of the moment, no one noticed that he had an infected wound on one of his fingers, the result of an accident during an operation. The infection developed into a general sepsis, and he died two weeks later under circumstances that were strangely reminiscent of the death of his friend Jakob Kolletschka.

After his death, the seminal discovery of Philipp Semmelweis rapidly fell into oblivion, where it would remain for the rest of the century until it was finally resurrected, not least by the efforts of that great surgeon and noble character, Joseph Lister. Why did his contemporaries not immediately realize the true importance of Semmelweis's work on the prevention of puerperal fever? Surely his unexpected and ill-advised departure from Vienna, that took his friends completely by surprise, had something to do with it. If he had stayed on and with their help doggedly continued his fight against his hidebound adversaries, he might well have prevailed in the end. At the same time, there were deeper reasons for his misfortunes. It must be realized that at this time the struggle between contagionists and miasmatics was still going on, and there was no understanding of the role of bacteria in either surgical or puerperal fever. As we have seen, Semmelweis himself thought in terms of a poison that caused the fever and could be destroyed by chlorinated lime; the idea of sterility was alien to him. Consequently, there was no intellectually satisfactory explanation of the phenomenon he had discovered and the technique he had introduced. It would take until the last decades of the century before the medical world had fully recognized the central importance of bacteria in wound infections and contagious diseases.

The Patient Quaker

The tragic figure of Semmelweis is one of the great and long neglected heroes of medicine, and his indefatigable struggle to rid the maternity wards of the ravagings of puerperal fever can indeed be seen as the dawn of a new era. However, it was a man of an entirely different mould who would lead the way to modern surgery. An unassuming and thoroughly sound Englishman, who in every respect was the antithesis of the passionate and somewhat unstable Semmelweis.

Joseph Lister (1827–1912).
Courtesy of Bonnierförlagen, Stockholm, Sweden.

Joseph Lister (1827–1912) was from a well-to-do Quaker family in Essex. His father was a wine merchant with strong scientific interests, who despite his poor schooling made important contributions to optics, in particular the construction of powerful achromatic lenses in compound microscopes. Lister was educated at a Quaker school with a curriculum that emphasized mathematics, natural sciences and modern languages. At the age of 17, he entered medical school at University College, London, where he was a very conscientious and successful, if somewhat reticent student. During the clinical part of his studies, he took an interest in surgery, and after medical school he decided to spend a few months with James Syme, the most celebrated surgeon in Scotland, known not only for the incredible speed of his amputations but also for his strictness and exacting demands.

When Lister arrived at Syme's clinic in Edinburgh he must have made a very good impression, for his new chief soon made him a supernumerary house surgeon and Lister became a frequent guest in his family. Instead of returning to London, he stayed on in Edinburgh and a year later he was made resident house surgeon. In 1856, Syme entrusted him with an even more important responsibility when he gave Lister his eldest daughter Agnes in marriage. Their married life lasted until Agnes' death in 1893; though childless, they were however obviously very happy together.

After seven successful years in Edinburgh, Lister was appointed professor of surgery at the University of Glasgow. When he took charge of the surgical wards of the Royal Infirmary there, the mortality in surgical fever was as appalling as in most other major hospitals in Europe. The prevailing theory among surgeons was that the fever was caused by a process of putrefaction in the wound, a kind of combustion of the tissue in the presence of oxygen. A number of more or less airtight dressings were therefore employed to prevent the atmospheric oxygen from entering the wound. Lister, however, found it difficult to believe that oxygen caused the putrefaction and when the professor of chemistry, Thomas Anderson, drew his attention to Pasteur's work that demonstrated the impossibility of spontaneous generation and instead implicated the microorganisms in the air as the cause of the fermentation of exposed body fluids, the truth dawned on Lister. If these fluids could be infected by airborne microorganisms in Pasteur's experiments, surely such organisms might also give rise to infections in wounds and thereby cause surgical fever.

He began to look for a suitable substance that could be used to treat infected wounds and he immediately hit on a very effective disinfectant. Lister was aware that carbolic acid (phenol) had been successfully used to sanitize sewage and he decided to try it on wounds. Compound fractures, where the skin had been penetrated, were known to have a dreadful mortality and he chose these almost hopeless cases for his first experiments with the carbolic acid treatment. He cleansed the wound with a 5% solution of the disinfectant and then used dressings soaked in carbolic acid to protect it from airborne microorganisms. In a period of about two years, he treated 11 cases in this way and nine of those recovered. This almost incredible decrease in the mortality of compound fractures was the first spectacular triumph of the antiseptic principles.

When Lister published his results in *The Lancet* in 1867 under the heading *On a New Method of Treating Compound Fractures*, one might have expected his carbolic acid treatment to spread like wildfire all over the world, but not at all. He had quite a few admiring proselytes, especially among his immediate collaborators, but in many quarters his new method was received with indifference and scepticism. Particularly at the great hospitals in London, surgeons were generally unimpressed, but Lister found powerful opponents even closer to home. In Edinburgh, Sir James Simpson, famous for having introduced chloroform as an anesthetic, attacked him bitterly and not only belittled the importance of antisepsis but also insinuated that Lister had actually plagiarized the results of one Jules Lemaire, a pharmaceutical chemist in Paris — an entirely unjustified accusation. Lister vigorously defended himself against these allegations, but Simpson remained hostile and there were many other surgeons that shared his views.

In the meantime, Lister showed that carbolic acid was effective not only against infection of compound fractures, it also reduced the mortality of amputations in his own clinic from 43% to a mere 15%. At a meeting of the British Medical Association in 1867, he also claimed that his wards were now free of surgical fever. Lister undoubtedly greatly exaggerated the importance of airborne microorganisms in wound infection. To combat this largely imaginary threat, he constructed various types of sprays that vaporized dilute solutions of carbolic acid in the operation theater. At this time, the toxicity of phenol was not realized, and there is no doubt that the lavish use of this compound posed a threat to those constantly exposed to it, not least when inhaled as a vapor. When Lister became aware of its dangers and his own exaggerating of the risk of airborne infections, he finally discontinued altogether the use of carbolic acid sprays. He also tried to minimize the direct contact between the phenol and the skin of the patients by using different kinds of protective dressings.

In 1869 his father-in-law, James Syme, suffered a stroke and Lister was appointed as his successor in Edinburgh. Soon afterwards his father died, and the next year both Syme and his old antagonist, Sir James Simpson, passed away. In spite of the sad beginning of his years as professor in Edinburgh, this was perhaps the happiest period in Lister's life. He continued his work on antisepsis and also

Theodor Billroth (1829–1894).
Courtesy of Bonnierförlagen, Stockholm, Sweden.

introduced a number of technical improvements in surgery, work that he had already begun in Glasgow. His fame as a renewer of surgery continued to grow, particularly abroad in countries like France and Germany. His English colleagues were not always as enthusiastic, and when he transferred to London in 1877 to become professor of surgery at King's College, the reception was distinctly cool. The other London surgeons were uninterested, and both students and nurses were obstructive and protested against the laborious methods of antisepsis that Lister introduced in the clinic, but by patient perseverance he overcame all resistance.

When the recognition came it was indeed overwhelming, and he was showered with honors both at home and abroad. He was made a baronet in 1883, and in 1897 he was raised to the peerage, Prussia conferred the Ordre pour le Mérite on him and there was a deluge of honorary doctorates. He received the freedom of Glasgow, Edinburgh and London, and from 1895–1900 he was president of the Royal Society. All these honors were indeed well deserved, but one cannot help thinking about Ignaz Semmelweis, the great and unsung pioneer whom no one seemed to remember, while Pasteur and Lister were celebrated all over the world. Characteristically enough, Lister — a great man also in a moral sense — was the first to acknowledge the importance of the forgotten pioneer, when he became aware of Semmelweis, albeit relatively late, and his lonely fight against puerperal fever. In Lister's own words, when he stood on the peak of his fame and was generally recognized as the father of modern surgery: "To this great son of Hungary surgery owes most."

A Triumphant Progress

During the second half of the 19th century, surgery started its triumphant progress towards a position as the dominating clinical field, perhaps the only one where the physicians could actually offer their patients a really effective treatment. The introduction of general anesthesia and the principles of antisepsis had liberated surgery from the shackles of pain and wound infection, and it was now free to embark on a development that would previously have been impossible. During this new era of surgery, that the work of Lister had ushered in, many of the pioneering contributions were made in the German-speaking countries.

The history of medicine contains quite a few infant prodigies, but some of its greatest names, like Claude Bernard and Louis Pasteur, were considered more or less hopeless during their schooldays. To this list of stupid schoolboys, later turned geniuses, can be added the name of the great German surgeon Theodor Billroth (1829–1894). He was the son of a clergyman on the island of Rügen in the Baltic and came from a music-loving family, an inclination that was very obvious also in the young Theodor. When he was five years old, the family, upon the death of his father, moved to the ancient university town of Greifswald. Here, Billroth attended the Gymnasium, where he was seen as a rather mediocre pupil with considerable difficulties in regard to both languages and mathematics, and no particular talents except for music.

When he graduated from the Gymnasium, friends of the family, who were professors at the medical faculty of the local university, persuaded him to enter medical school. The incentive seems to have been purely economic; they convinced young Theodor that medicine offered a fairly certain way of making a living. He began his studies in Greifswald and Göttingen, but in 1851 he moved to Berlin where he came under the influence of some of the leading figures in German medicine, for instance the great physiologist Johannes Müller, the famous clinician Johann Lucas Schönlein and the surgeon Bernhard von Langenbeck, whose assistant he became in 1853, after having graduated and passed his state medical examination. During his Berlin years, he was scientifically active mainly in pathological anatomy, where he made a respected name for himself. In Berlin, Billroth also met his future wife Christine Michaelis, daughter of the court physician, and they were married in 1858.

So far, there was nothing in Billroth's achievements that portended the distinguished surgical career that lay ahead of him. He was mainly appreciated as a pathologist and had, in fact, been seriously considered for the chair in Berlin, that finally went to another pathologist of even greater renown — Rudolf Virchow. The turning point was his appointment as professor of surgery in Zürich in 1860. At this time, the development of general anesthesia had greatly encouraged surgeons to attempt new and daring operations, but wound infections still made the mortality of all major operations appallingly high. Billroth had always been interested in this problem and he had formed a theory of his own. He assumed that all surgical fever was caused by a bacterium that he called *Coccobacteria septica* and that according to Billroth produced a kind of ferment — *phlogistisches Zymoid* ("inflammatory ferment") that was the immediate cause of inflammation and fever. This can perhaps be seen as a step in the right direction, but unlike Lister he was never able to device an effective method of prevention on the basis of his theory. Later, he became one of the first to adopt Lister's antiseptic technique, although he was rather worried by cases of phenol poisoning that he observed. This problem was solved when, particularly in Germany, surgeons switched to aseptic methods where phenol was dispensed with in favor of sterilization by heat of everything that came into contact with the wound.

During the Zürich years, Billroth laid the foundation in skill and experience that would later enable him to perform the daring operations that opened up new frontiers in surgery. He also made an important contribution to the statistical analysis of surgical results by insisting on absolute honesty in the reporting of data. This may seem self-evident to us, but was in reality something of an innovation and a break with the old tradition of reporting sensational and fortunate results and suppressing the failures. After all, why bother with dead or mutilated patients, when one could report much more interesting successes instead? In his pioneering *Yearly Report from the Surgical Clinic in Zürich*, Billroth faithfully reported all the facts, without any attempt at embellishing the truth, and he brought this principle of absolute truthfulness with him, when in 1867 he was appointed to the prestigious chair in surgery at the university of Vienna.

Billroth had already declined offers from Rostock and Heidelberg, but perhaps he was particularly attracted to Vienna because of the exceptionally rich musical life there. He became a close friend of Johannes Brahms, who dedicated two of his string quartets to Billroth. His home in Vienna was a haven for music lovers where the host himself played the violin. Another example of his devotion to music is his book *Who is Musical?* that he worked on during his last illness. Perhaps a somewhat unexpected item in the bibliography of a great surgeon, but very illuminating when it comes to understanding Billroth, the man.

Billroth certainly had the gift of friendship with his upright honesty, deep concern for the well-being of his fellow creatures and almost childlike delight in the simple joys of life. As might have been expected, he attracted a great number of talented disciples, the famous Billroth school, that dominated surgery in Central Europe in the decades around the turn of the century. He worked closely with his devoted assistants to lay the necessary foundation for each daring new operation by careful experiments on laboratory animals. In 1872, he made the first resection of the oesophagus, the next year he performed a total laryngectomy and in 1881 he successfully removed a cancerous pylorus. The last operation involved the resection of a considerable part of the stomach and there were a number of worrying questions that had to be answered by animal experiments before it could be attempted.

This careful, methodical experimental groundwork is characteristic of Billroth and goes a long way to explain his greatness as a surgeon. At the same time, he always maintained that great scientists had a strain of the artist in them. In a letter to Brahms in 1886, he says that the great scientist "is a kind of artist, highly imaginative and with a childlike nature. That brings me back to my hobbyhorse. Science and the arts spring from the same well." It is easy to understand that a man like Billroth inspired not only admiration but also love among his many followers.

During his last years, Billroth suffered from a cardiac condition that seriously weakened his previously robust health. At the end, he tried to take stock of his life and evaluate his achievements. One might have expected him to dwell on some of his pioneering operations and the opening up of the field of abdominal surgery, but it is his role as an inspiring teacher and the founder of a school that he values most. This may perhaps seem strange, but he is not the only leading figure in medicine, who has arrived at the same conclusion when faced with this question. Science is after all not the pursuit of the lonely wolfe; rather it is a very social activity that brings people close together in order to solve a common problem. Perhaps the research team can be seen as akin to an orchestra, whose members make music together; a simile that the friend of Brahms might have appreciated.

ROBERT KOCH

The Spare Time of a District Medical Officer

Koch came from a family of mining officials in Harz where he was born on 11 December 1843 in the little town of Clausthal. His father had, by talent and energy, advanced from common miner to a leading position in the mining industry where he eventually had the title of *Geheimer Bergrat* with responsibility for all mines in the district. This meant that in later years he had a fairly decent income. He could certainly use the money, since in his marriage to Mathilde Biewend, the daughter of a mining official, he had 13 children to provide for. Two of the children died in infancy, not an unusual proportion in those days, but Mathilde must nevertheless have had rather burdensome duties as a housewife with a family that size. Her husband was much too preoccupied with his responsibilities as head of the mining operations to take a more active part in the upbringing of the children. It is the more admirable that she nevertheless found time to encourage Robert's interest in nature. In this, her brother Eduard Biewend, a man of considerable education and knowledge, helped her by taking the boy in hand and awakening his interest in the natural sciences. Together, they made excursions to collect plants, insects and minerals that Robert then classified with the aid of a magnifying glass. Furthermore, his uncle also introduced him to photography, something that would later on be of great importance for Koch's research.

He seems to have done tolerably well at school, although for some reason he had to spend two years in one of the classes. In 1862, he graduated from the Gymnasium with decent characters, except in religion, Latin, Greek, Hebrew and French where he just had passing grades. Strangely enough, the young student had taken it into his head that he should go on to study linguistics. Fortunately, the headmaster managed to convince him that his talents were more in the direction of natural sciences, and Koch instead began to study botany, physics and mathematics at the University of Göttingen. The prospect of becoming a school teacher did not appeal to him, and after two terms he went over to the medical faculty. At this time, Jakob Henle was one of the leading professors there and as we know he had already in 1840 suggested that living, invisible organisms were the cause of contagious diseases. One would assume that the early contact with Henle and his ideas must have been important for the young Koch. However, there was no teaching of bacteriology at Göttingen while Koch was a student there. Even so, Pasteur's demonstration of the impossibility of a spontaneous generation of life must certainly have been discussed there, and Koch was probably early on made aware of the existence of a new fascinating field of research — microbiology. Already during his lifetime it would revolutionize medicine, not least through his own efforts.

Robert Koch (1843–1910).
Courtesy of the Nobel Foundation, Stockholm, Sweden.

Koch got his MD and was licensed to practice medicine in 1866, and already the next year he married Emmy Fraatz, the daughter of a high official in Clausthal. At least to begin with, their marriage seemed a happy one, and in 1868 Koch's only child was born, a daughter with whom he had a warm relationship throughout his life, even after the breakdown of his first marriage. In the beginning, married life was full of economic difficulties and frequent moves to new places in order to better his incomes as a doctor in private practice. After a period as a military surgeon during the war with France in 1870/71, he returned to civil practice, and passed the examination for a position as Kreisphysikus (district medical officer). On the recommendation of an influential patient in 1872, he was appointed Kreisphysikus in the little town of Wollstein, in what was then the Prussian province of Posen (now Poznán in Poland). In this rural idyll, where Koch became a highly respected and popular country doctor, the little family spent eight apparently happy years. It was here that he began his remarkable career as the leading bacteriologist in the world together with Pasteur.

Nowadays, we tend to emphasize the crucial importance of the intellectual milieu for a young scientist's rise to eminence, his or her need for a mentor to emulate and the inspiration and strength that comes from being a member of a closely knit, well-functioning research group. This is a good argument and we can sustain it by pointing to veritable genealogies in science that time and again prove how excellence breeds excellence. At the same time, we should not lose sight of the many striking exceptions to the rule and Robert Koch is a case in point here. It is hard to imagine a less stimulating environment and one less conductive to experimental medical research than the rural backwater of Wollstein. The only thing that could be said to favor his research efforts was the fact that Koch's practice among the local residents seems to have left him some spare time.

Perhaps it was the agricultural surroundings that made Koch become interested in anthrax even before Pasteur had begun his studies of the disease. What fascinated Koch was the observation that the big, rodlike, easily recognizable anthrax bacillus, first described by the French pathologist Casimir Joseph Davaine, sometimes seemed to be absent from the blood of animals that had died of anthrax. Nevertheless, this apparently bacillus-free blood could still cause anthrax in healthy animals. It seemed as if the anthrax bacillus could exist in an alternative, less conspicuous form that was also infectious.

In the early days of bacteriology, it was widely believed that bacteria did not have a definite morphology but could, for instance, sometimes appear as a rod, sometimes as a sphere. The phenomenon was called pleomorphism, and the well-known botanist Karl von Nägeli (1817–1891) as late as 1878 declared:

"There are no true bacterial species. On the contrary, the variability of bacteria is unlimited. The same species might, during generations of growth, assume different morphological and physiological forms which over the years sometimes would cause milk to turn sour or protein to putrefy, sometimes cause diphteria or typhus fever, cholera or recurrent fever." Against this unlimited belief in the ability of bacteria to change morphology and character, another great German botanist, Ferdinand Cohn (1828–1891), maintained that bacteria could be classified according to their morphology in the same way as plants. A spheric bacterium continued generation after generation to show the same spherical form and never assumed the form of a rod.

In his rural surgery, Koch had established a laboratory behind a curtain and here he conducted his research. He had a good quality microscope, which he had fitted with a simple apparatus that allowed him to keep an object, for instance a drop of liquid medium, at a constant temperature for days on end. He also had equipment for microphotography with a small closet for a darkroom. His wife seems to have functioned as unsalaried laboratory assistant during this period. One might wonder if she was as enthusiastic as Koch himself about his research efforts.

In this simple laboratory, Koch grew anthrax bacteria in droplets of different media under varying conditions. He could then follow the transformation of the big, rodlike bacterium into a much smaller spore and also the reverse process by which the spore gave rise to a rodlike bacterium. The discoverer of the anthrax bacterium, Davaine, had never observed this tranformation, but Ferdinand Cohn had in fact predicted that the anthrax bacillus might have the ability to form spores. Obviously, the spore represented an alternative form of the bacterium, one that was more resistant to unfavorable conditions, like for instance prolonged exposure to drying. Koch realized that this adaptation of the bacterium to different environments and growth conditions was something entirely different from the fantastic transformations that Nägeli and other pleomorphists had suggested.

In the spring of 1876, Koch was ready to demonstrate his discoveries to Ferdinand Cohn in Breslau and the great botanist was sufficiently impressed by Koch's work to accept it for publication in his own journal *Beiträge zur Biologie der Pflanzen* ("*Contributions to the Biology of Plants*"). That is the reason why one of the pioneering publications in bacteriology, Koch's seminal *Die Ätiologie der Milzbrandkrankheit, begründet auf die Entwicklungsgeschichte des Bacillus anthracis* ("*The Etiology of Anthrax, Based on the Ontogeny of the Bacillus anthracis*"), appears in a botanical journal. At the same time, it illustrates the close relationship between botany and the young science bacteriology. Jakob Henle was already of the opinion that his hypothetical microorganisms belonged to the plant kingdom.

The almost explosive development of bacteriology during the latter decades of the 19th century depended to a large extent on two simple but exceedingly important technical improvements that were mainly due to Koch. The German pathologist Carl Weigert (1845–1904) had pioneered the use of synthetic aniline dyes to stain bacteria. Koch improved Weigert's procedure by the simple expedient of drying thin layers of bacteria on glass slides so as to fix them to the glass surface before staining. They could then be inspected and photographed in the microscope. The pictures that he obtained using this technique were so sharp and detailed that they could, for instance, be used to characterize and classify the bacteria found in infected wounds. When Koch demonstrated his results to one of Germany's leading pathologists, Julius Cohnheim (1839–1884), and a group of scientists that included his future collaborator and close friend, Paul Ehrlich, they were received with great interest and appreciation. On the other hand, a later visit to the Nestor of German pathology, Rudolf Virchow in Berlin, was not as successful. Virchow was not inclined to accept the role of microorganisms in contagious diseases and instead entertained views that were almost miasmatic. In coming years, Koch and Virchow would often be on collision paths with each other.

Koch was a strong adherent of Cohn's theory that bacterial species had characteristic morphological and physiological properties that remained constant for countless generations, and in 1878 he became involved in a heated discussion with Karl von Nägeli who had expounded his pleomorphistic ideas in a recent book. Koch's tendency to assert his opinions in a clear and unequivocal, not to say brutal way, would time and again lead to conflicts with other scientists, among them also one of his most outstanding collaborators, Emil von Behring. The same year that Koch had quarrelled with Nägeli, he published his important study of wound infections: *Untersuchungen über die Ätiologie der Wundinfektionskrankheiten* ("*Studies of the Etiology of Wound Infections*"). This had for a long time been a contested question even after the majority of physicians had accepted that wound infections were caused by bacteria. Joseph Lister had been inclined to pleomorphistic ideas and the same could be said of Theodor Billroth with his hypothesis about all wound infections being caused by the same bacterium, *Coccobacteria septica*.

Using tissue samples from infected wounds, Koch was able to transmit the infection to laboratory animals. When he analyzed the experimental infections, that eventually killed the animals, he could demonstrate a number of morphologically different bacterial species. He concluded that wound infections could be caused by a variety of bacteria belonging to genetically different species, to use a modern terminology. Koch's arguments convinced both Lister and Billroth, who from now on supported his theory of specific bacterial species causing defined

Julius Cohnheim (1839–1884).
Courtesy of Bonnierförlagen, Stockholm, Sweden.

diseases. Another important result of Koch's work with wound infections were his famous postulates, that had in fact been earlier suggested by his old teacher Jakob Henle. The great difference is of course that while Henle based his postulates on purely theoretical considerations, Koch could present experimental evidence to sustain his theses. Koch's postulates can be summarized approximately as follows:

1. It must be possible to demonstrate the bacterium (or virus), believed to be the cause of a certain illness, in all diseased organisms.
2. The suspected microorganism must be isolated in pure culture.
3. Using the pure culture it must be possible to infect suitable laboratory animals with the disease.

The Great Anthrax Quarrel

Koch's success as a part-time bacteriologist in leisure moments, pilfered from his medical practice in Wollstein, had been incredible, but now it became increasingly clear that entirely different recourses were needed in order to realize his scientific ambitions. What had become necessary was a well-equipped laboratory with highly trained technical staff and the opportunity for Koch to surround himself with a group of young, enthusiastic scientists, and last but not least, the possibility to devote all his time to research. Ferdinand Cohn had tried to create a professorship for Koch in Breslau, but the effort had failed and it was not until 1880 that on the recommendation of Julius Cohnheim he was appointed *Regierungsrat* (a permanent senior position) at the Imperial Ministry of Health in Berlin. His laboratory was housed in what had previously been an apartment building close to the Charité hospital, and here Koch started with two young military doctors, Friedrich Loeffler and Georg Gaffky as assistants, the first members of what would eventually become the leading bacteriological research group in the world.

His boss, Friedrich Struck, gave Koch what in reality amounted to a free hand in the choice of research projects and only a year later he published another pioneering improvement of bacteriological technique, the growing of bacteria on solid substrates. To begin with, he had suspended bacteria in a liquid medium containing gelatin that was then allowed to solidify, but soon he turned to growing bacteria on the surface of solid media. He now had a simple and reliable method for obtaining bacterial colonies that had grown from a single bacterium, something that had been both laborious and unreliable with previous, rather primitive dilution techniques. When Koch in 1881 demonstrated his new technique during a visit to Lister's laboratory in London, Pasteur was also present and he immediately declared the growth on solid media to be *un grand progrès* ("a great progress"). Unfortunately, it would not be long before the two giants of bacteriology quarrelled violently with each other.

The reasons for the conflict were complicated and partly had to do with their personalities that seem to have been incompatible. To some extent, the confrontation might have been exacerbated by national antagonism after the war of 1870/71 that had ended in a catastrophe for France. Pasteur was a great patriot with a streak of chauvinism and the same could probably be said of Koch. However, there were more obvious reasons. All the time since he began to take a serious interest in anthrax in 1877, Pasteur had been inclined to claim for himself the credit for having discovered that certain bacteria can form spores. He based this on some rather diffuse observations that he was supposed to have made while working with the sick silkworms in the 1860s. In reality, there is no doubt that it was Koch's mentor and patron, Ferdinand Cohn, who in 1875 had discovered the phenomenon of spore formation as part of the life cycle of *Bacillus subtilis*.

What made the conflict really flare up was the publication where Koch described his method of growing bacteria on solid substrates and other technical improvements. He also took the opportunity to attack vehemently the general quality and reliability of Pasteur's bacteriological work. Koch emphasized that the liquid media which Pasteur used did not, unlike growth on solid substrates, give pure cultures. In particular, he criticized Pasteur's attenuated anthrax bacteria obtained at 42°C and asserted that it was just a question of contaminated cultures. On the other hand, this allegation was not consistent with Pasteur's successful vaccinations with precisely these temperature attenuated cultures and Koch actually maintained for years that the documentation of Pasteur's vaccination success against anthrax was unsatisfactory, not to say fraudulent.

Pasteur was not a person whom you attacked with impunity, and when he became aware of Koch's accusations he mounted a counterattack on a congress in Geneva that they both attended. In his speech, he dismissed the German group as inexperienced, pointed to his successful vaccinations in recent years and challenged Koch to answer him. Koch, however, bided his time and it took three months before he published a rebuttal to Pasteur's statement. He now, somewhat unexpectedly, accepted that anthrax bacteria might be obtained in a temperature attenuated form, but gave some of the credit for this discovery to the French veterinarian Toussaint. In 1880, Toussaint had used anthrax-infected blood, which had been heated to 55°C, for vaccination purposes and had in some cases observed a certain immunization. Koch then proceeded to criticize the results of Pasteur's own vaccinations, based on some negative results that had been reported by the Veterinary School in Turin. He also repeated his previous accusations against Pasteur for working with impure cultures, caused by faulty technique, and what was more, he questioned Pasteur's medical and bacteriological knowledge in general.

Pasteur's answer to this took the form of an open letter to Koch where he sarcastically and with some justification represented Koch's paper as a somewhat reluctant admission that Pasteur had after all been right about the attenuated anthrax bacteria. The next year a comprehensive and successful immunization program using Pasteur's vaccine was carried out all over Europe, including Prussia. Nevertheless, when Pasteur in 1887 triumphantly declared that now even the "Berlin school," as he called them, were convinced, there was an immediate answer from Koch to the effect that he had in no way changed his opinion about Pasteur's anthrax vaccine. It is hard to see this as anything but a manifestation of Koch's well-known stubbornness, in many ways certainly an asset, but in this particular case more an example of his inability to admit having been wrong.

There was undoubtedly a marked difference between the schools of Pasteur and Koch in terms of the general direction of their research. With his strong wish to see his work result in practical achievements that could benefit humanity and above all his beloved France, it was natural that Pasteur should emphasize preventive measures like vaccinations. At this time, real treatment of infectious diseases was out of the question, it would take almost a century before we had effective drugs against bacterial infections. Koch, on the other hand, with his thorough medical and morphological training, had pioneered new methods for growing bacteria and obtaining them in pure culture so that they could be identified and related to different clinical conditions. He was also less inclined than Pasteur to accept that bacteria, through variations in their growth conditions, could be attenuated in terms of their ability to cause disease (virulence) even if they looked morphologically the same. Thus, his tendency to dismiss Pasteur's attenuated anthrax bacteria as merely impure cultures. Pasteur, on the other hand, maintained rightly that his attenuated bacteria represented real changes in the properties of microorganisms, for instance virulence, but that the virulence could be recovered by passage through suitable hosts. This difference in the general attitude of French and German bacteriology also explains why it was mainly in Germany that great progress was made in terms of the isolation and characterization of the pathogenic bacteria that caused, for instance, tuberculosis, cholera, diphteria, gonorrhoea, etc.

Théophile Laennec (1781–1826).
Courtesy of Bonnierförlagen, Stockholm, Sweden.

Tuberculosis, An Infectious Disease

The French clinician and pathologist Théophile Laennec (1781–1826) had been particularly interested in tuberculosis of the lungs, not least because he himself suffered from this condition. In spite of his keen interest, he never realized that tuberculosis was an infectious disease. Laennec was not the only doctor to make this mistake. The majority of physicians, even as late as the 1880s took the same view. Fracastoro's perceptive conclusion 300 years earlier had long ago been forgotten. It was not that physicians at the time were unaware of the sufferings caused by tuberculosis. On the contrary, it was generally recognized that this was the main reason for sickness and death in the young and middle-aged, a scourge that claimed innumerable victims in all classes of society. Most physicians were of the opinion that it was mainly a question of a constitutional, inherited tendency to develop the disease, even if bad housing conditions and malnutrition undoubtedly contributed. In the spirit of Galen, one talked about a dyscrasia, i.e. a disturbance of the body humors; in any case it was not infectious.

However, experimental data were forthcoming which suggested that tuberculosis was not a dyscrasia at all but rather an infectious disease, exactly as Fracastoro had claimed. The French physician Jean Antoine Villemin (1827–1892) succeeded in transmitting cattle tuberculosis to rabbits in 1865, and some years later Julius Cohnheim showed that tuberculous material inoculated in the anterior chamber of a rabbit's eye gave rise to tuberculous lesions of the retina. Such results were of course encouraging for doctors who believed in tuberculosis being of infectious origin, but it still remained to produce the definite proof of what was so far only a hypothesis. It would require Koch's masterly command of the bacteriological technique to obtain the elusive *Bacillus tuberculosis* in pure culture and elucidate its role as the cause of the disease.

The tubercle bacillus turned out to have properties that made it considerably more difficult to study than the anthrax bacterium. It was much smaller and particularly hard to stain. After many attempts, Koch was able to demonstrate the bacillus in tuberculous tissues that he had stained with alkaline methylene blue. The staining procedure was soon improved by Paul Ehrlich who also suggested that the tubercle bacillus was surrounded by a waxy coat, an idea that Koch eventually accepted. A great problem was that the bacillus did not grow on substrates that gave a luxuriant growth of anthrax bacteria. Finally, Koch succeeded in growing the tubercle bacillus on the sloping surface of heat coagulated serum. However, it was a very slow process and it took two weeks before any bacterial growth could be observed. Using this technique for growing and staining, Koch could then demonstrate the presence of the tubercle bacillus both in the typical tubercular changes in tuberculosis of the lung and in tuberculous material from other organs. This definitely put an end to the old idea that it was a question of two different illnesses, a mistake that already Laennec had rejected, while Virchow still stubbornly adhered to it. The demonstration of the tubercle bacillus in the sputum of the patient became a diagnostically extremely important test, which also made it possible to decide if the patient was in a contagious stage of the disease. Furthermore, pure cultures of the bacillus could transmit tuberculosis to a number of different laboratory animals, thus meeting the requirements of Koch's and Henle's three postulates.

In a talk with the title *Über Tuberculose* that he gave in the Physiological Society of Berlin on 24 March 1882, Koch described his results before an audience that completely filled the lecture hall, and where also his future collaborator Paul Ehrlich was present. Ehrlich would later in life describe this as his greatest scientific experience. Koch's faithful pupil Friedrich Loeffler was of course also there and 25 years later he recalled what was surely one of the greatest moments in the history of bacteriology.

It would seem that all present agreed that Koch was no great orator. On the contrary, according to Loeffler he spoke slowly and seemed to fumble for words. Nevertheless, Koch made a great impression on the audience by the stringency and logic of his unadorned presentation. Loeffler especially points out that there was no discussion after the talk, a completely unique occurrence in the history of the Society. His own explanation was that Koch had completely convinced everyone present. A couple of weeks later, Koch published a paper in *Berliner klinische Wochenschrift* (*"The Weekly Clinical Journal of Berlin"*) with the title *Die Ätiologie der Tuberculose* (*"The Etiology of Tuberculosis"*), one of the real landmarks in medical literature. The fact that it contains certain mistakes, for instance Koch did not distinguish between human and bovine tuberculosis but considered them to be caused by the same bacillus, cannot detract from our high opinion of it. When Koch later on became aware of his mistake, he went one step too far in the opposite direction and claimed that in practice, one could disregard the possibility of bovine tuberculosis being transmitted to humans. This would lead to scientific conflicts with both Emil von Behring and Lord Lister.

Koch showed no mercy to the ignorant obscurants who, in spite of overwhelming proof, continued to doubt that the tubercle bacillus was the cause of tuberculosis. He was brutally straightforward when he took them to task for faulty technique and general ignorance. Even if Virchow and his followers remained sceptical about the role in general of bacteria as the cause of disease, Koch can be said to have been victorious all along the line. His scientific eminence was also recognized by the Prussian authorities who awarded him the distinguished rank of *Geheimer Regierungsrat*. His powers would soon be claimed by other, global problems in bacteriology and hygiene.

The Cholera Vibrio

In 1883, the cholera had come to Egypt from India travelling with the Islamic pilgrims who undertook the journey to Mecca demanded by their religious faith. Pasteur had warned the French government that this outbreak so close to the Mediterranean coast of France would pose a threat to the country itself as well as to the rest of Europe. On his advise, the government sent a medical commission to Egypt, which included several of Pasteur's pupils, among them the young Louis Thuillier, one of the tragic victims of what might perhaps be called the heroic period of bacteriology. The German government, not wanting to be outdone by France, sent a commission of its own to Egypt with Koch at its head.

From the very beginning, the French bacteriologists had been haunted by misfortunes. Their primitive technique, trying to obtain pure cultures of the pathogenic bacterium in liquid media, was a failure and resulted only in a confusing mixture of intestinal bacteria. To begin with, great importance was attached to particles found in the patients' blood, which later turned out to be just ordinary blood platelets (thrombocytes), normal components of the coagulation system. During these futile efforts, Thuillier was infected with cholera. When Koch learned of Thuillier's illness, he visited his young colleague and competitor, who was then at the point of death. Thuillier supposedly asked Koch if what the French group had found in the blood of the patients was really the cholera bacterium. Against his better judgement, Koch should then have answered the dying youth in the affirmative. This is the kind of moving tale that one is naturally inclined to distrust. On the other hand, it is consistent with a side of Koch's personality, which he did not always exhibit to his French colleagues — a warm and humane trait that undoubtedly existed under the somewhat gruff surface.

With their superior technique of growing bacteria on solid media, Koch and his collaborators were able to isolate a small, comma-like bacterium that was found in abundance both in the content of the small intestines and in the intestinal wall itself of cholera patients. At the time of Thuillier's death, the epidemic was already disappearing from Egypt, and Koch therefore moved to Bengal in India where cholera was endemic. Here, they could confirm their Egyptian findings and demonstrate the presence of the "comma bacterium," as it was now beginning to be called, in as many as 70 cholera cases. A difficulty was that pure cultures of the comma bacterium did not cause cholera when administered to a number of laboratory animals, such as monkey, dog, chicken and mouse. However, Koch could later show that the bacterium, which we now call *Vibrio cholerae*, could cause cholera in guinea pigs when introduced directly into the stomach in an alkaline medium. Another proof was the unintentional infection of human guinea pigs in the form af laboratory staff accidentally exposed to pure cultures of *V. cholerae*, who were taken ill with the disease. In his last report from India to the German Imperial Department of Health in March 1884, Koch pointed to the village ponds of Bengal, from which water was drawn for the entire village population, as the main source of infection. By isolating the *V. cholerae* from the drinking water of a cholera-stricken village, he had in fact experimentally verified the conclusions of John Snow in London 35 years earlier.

The members of the committee were welcomed as heroes when they returned to Germany. The emperor gave Koch a distinguished order, the Reichtag voted him a reward of 100,000 marks and the outstanding surgeon Ernst von Bergmann paid homage to him at a great banquet given in his honor in Berlin. However, all Koch's colleagues were not that enthusiastic. At two conferences summoned on his initiative, the comma bacterium was discussed under the chairmanship of Virchow, no friend of Koch or bacteriology in general. The leading German hygienist, Max von Pettenkofer, spoke at the last conference and questioned the role of the comma bacterium as the cause of cholera. Instead, he indulged in what was almost miasmatic speculations. Koch of course vigorously defended his findings and their clash led to a long and bitter conflict.

Even a country as medically advanced as Germany was not safe from the ravages of cholera at the end of the 19th century, in spite of the fact that both the cause of the disease and the way it was disseminated had been elucidated, although it had not been generally accepted yet. In August 1892, a serious epidemic broke out in Hamburg with 18,000 cases and a mortality of 45%. The authorities turned to Koch for advise on how to contain the epidemic, and he recommended such in our view self-evident precautions as early diagnosis using bacteriological methods, disinfection of contagious excreta, and last but not least, a strict water hygiene. Incredibly enough, these measures were ridiculed by his antagonist Max von Pettenkofer. In order to emphasize his repudiation of Koch's theory of the comma bacterium as the cause of cholera, Pettenkofer and one of his followers, professor Rudolf Emmerich, were foolhardy enough to swallow some fresh cultures of *V. cholerae*. Of the two human guinea pigs, Emmerich sure enough came down with cholera, but Pettenkofer himself escaped with just a diarrhoea. Since neither of the participants in the dangerous experiment had died, Pettenkofer was of the opinion that he had proved his point — cholera was not caused by Koch's comma bacterium. Koch was content with writing an extensive report about how the epidemic in Hamburg had been combatted using his methods that were soon generally accepted both in Germany and internationally.

A Regrettable Mistake

In 1885, he was appointed to a newly created chair in hygiene at the Berlin University and he was surrounded by outstanding collaborators who had come from all over the world to learn the new science, bacteriology. In addition to Loeffler and Gaffky, who had been with him from the beginning, Paul Ehrlich, Emil von Behring and Shibasaburo Kitasato later joined the group, just to mention a few of the most famous names. In 1891, Koch became director of the new Institute for Infectious Diseases in Berlin and he remained in this position until 1904. It almost seemed as if nothing he undertook could ever fail. Nevertheless, he was on the verge of becoming involved in the gravest of all the conflicts that succeeded each other in the life of this pugnacious scientist.

At the end of the 1880s, Koch's own research became increasingly preoccupied with attempts to find an effective remedy against tuberculosis. He worked mostly alone, surrounded by a certain secrecy. At an international medical congress in Berlin in August 1890, Koch unexpectedly announced, at the end of a fairly unexciting talk, that he had found a substance that prevented growth of tubercle bacilli both in test tubes and in laboratory animals. Guinea pigs injected with the substance became resistant to infection with tubercle bacilli, and in animals where tuberculosis had already become manifest the progress of the disease could be checked. In November the same year, there were sensational reports from doctors, closely associated with Koch, who claimed to have obtained spectacular results with the new remedy in the treatment of human tuberculosis. Koch himself pointed to the destructive effect of the substance on tuberculous tissues and warned against using it in advanced cases of tuberculosis of the lungs. On the other hand, he claimed at the same time that it was a certain remedy against early cases of tuberculosis.

The sensation was immense, both in the medical world and among the public. At this time, tuberculosis was very widespread in all strata of society. There was hardly anyone who did not have relatives and friends among the victims of the dreaded disease. Crowds of patients gathered in Berlin hoping to be cured by Koch. At the municipal Moabit Hospital, new wards with a total of 150 beds were opened to accommodate impecunious patients suffering from tuberculosis who were treated with Koch's unknown substance under the supervision of his collaborator Paul Ehrlich. To begin with, the results looked very promising and the optimistic reports were given additional credibility by the Prussian minister of Public Instruction, Gustav von Gossler, who loudly advertised the new remedy. It would seem that Gossler intended to make a public monopoly of it, and this was the reason why he advised Koch not to give any details about the mysterious substance and how it was obtained.

When the initial enthusiasm had subsided, there was increasing criticism. The idea to make a monopoly of the production was generally disapproved of and Koch was criticized for his refusal to give any details about the research that had led to the discovery of the active substance. Even worse, there was convincing evidence that the substance was actually toxic. As one might have expected, Koch's old antagonist Virchow had done his bit by demonstrating in his autopsies a strong inflammatory reaction where the mysterious medicine had been injected. He also emphasized the possibility of a generalization of the infection resulting in miliary tuberculosis. Demands that Koch should describe his preparation and how it was obtained became more and more pressing. When he finally did this in January 1891, it turned out to be just a glycerol extract of a pure culture of tubercle bacilli, that was all. No information was given about the experiments on guinea pigs that Koch had originally referred to.

It was now increasingly called in question if what later came to be known as tuberculin, had any therapeutic effect on tuberculosis at all. The criticism grew incessantly, and even such a loyal friend and supporter as Ernst von Bergmann began to have doubts. When he demanded to see the protocols from the autopsies that had supposedly been done on Koch's tuberculosis-infected guinea pigs, made resistant to tuberculosis by injection with tuberculin, it appeared that no autopsies had been performed on these animals. Koch's claims of both a preventive and a therapeutic effect of tuberculin in guinea pigs simply had no real basis. The scandal was now a fact and it had political consequences. Gustav von Gossler was held to be responsible for the sensational manner in which Koch's tuberculin was introduced, long before it had been properly tested scientifically. Koch himself was hardly blameless here, but Gossler was made the scapegoat and had to resign as minister.

Koch's enemies — with his aggressive and pugnacious manner of debate he had managed to make quite a few — now saw their chance. His old adversary, Rudolf Virchow, violently criticized the cost of the Institute for Infectious Diseases which was being created for Koch. His salary as director of the institute was much too high (20,000 marks per annum) and his research budget was comparable to that of the entire Berlin University, complained the leading figure of German pathology when he raised the question in the Reichtag. However, nothing came of this attack since Koch had support in high places and in the end he got both his institute and his money. Eventually, the allergic phenomenon which tuberculin causes when it is injected into the skin of patients who have gone through a tuberculous infection, would prove to have a diagnostic value. This can of course be considered as something of a rehabilitation of Koch's tuberculin, but his scientific reputation had, nevertheless, been severely damaged.

In addition to all these worries, his marriage was on the rocks. Koch and his wife seem to have been estranged for quite some time when he met a young woman, Hedwig Freiberg, who was 29 years his junior. When Koch first saw her, she was only 17 and he fell hopelessly in love with the beautiful and gifted girl. In his letters to her, he appears both in the role as passionate lover, assuring her of his ardent affection, and as the world-famous scientist who as a matter of course confided to her all the worries about his research. For instance, how his old antagonist Virchow sought to thwart him at every opportunity and the constant problems he had with financing the new institute. In one of his letters, written from the cholera-beset Hamburg, where he depicts the streets of the city as a battlefield covered with corpses, he expresses at the end of the letter his great joy at knowing that she is safe in the countryside. He also explicitly forbids her to eat anything that has not been thoroughly cooked, no fruit, no salad! The thought that he himself was in constant danger during his work in the stricken city obviously never entered his mind.

His wife Emmy agreed to a divorce when Koch transferred the ownership of his paternal home in Clausthal, which he had recently bought, to her and also otherwise made generous economic arrangements to provide for her. It appears that even after the divorce, she continued to take great interest in everything that concerned Robert Koch. In September 1893, Koch was married to Hedwig Freiberg, and as might have been expected this led to a number of complications. To make matters worse, she was an artist, a profession that did not enjoy much social esteem in imperial Germany. His divorce and subsequent remarriage was a scandal in society that certainly attracted much more attention than the unfortunate tuberculin business. People were very inclined to moralize in that golden age of bourgeois culture, before the First World War definitely turned everything in the old society upside down.

Global Bacteriologist

It is understandable if Koch became increasingly interested in all the research projects in what we now call the Third World, that he was constantly being invited to lead. After all, he was undoubtedly the greatest bacteriologist of his age, now that Pasteur's splendid career was drawing to its end. During the latter half of the 90s, he spent practically all his time abroad, mostly in Africa and Asia, working incessantly with a number of tropical diseases and most of the time accompanied by his young wife. What had to begin with looked like an odd marriage had turned out to be a very happy one. From South Africa came a request for help with a rinderpest that ravaged the cattle north of the Orange river. Koch with his scientific staff and of course also his wife, arrived in the district at the end of 1896 and immediately started his investigations. He soon established that it was not a question of a bacterial infection (the disease is caused by a virus) and attempts to obtain an attenuated sickness agent were unsuccessful. At long last, a short-lived immunity was achieved by using a mixture of serum from cattle, that had recovered from the illness, and infected blood from sick animals. Alternatively, livestock was inoculated with bile from animals that had recently died of the disease and eventually the rinderpest was brought under control.

Having got this far, the time had come to leave South Africa for India as the leader of a scientific commission dispatched by the German government in order to fight the epidemic of bubonic plague that had broken out in the northwest provinces. Alexandre Yersin (1863–1943) from the Pasteur Institute and Koch's own previous collaborator Kitasato had independent of each other discovered the plague bacillus, *Bacillus yersini*, in 1894. The commission, working together with scientists from several countries, now made epidemiological studies and confirmed the old observation that rats carried the disease, but did not realize the role of the flea as a vector for the plague bacillus. From India, Koch travelled to Dar-es-Salaam in what was then German East Africa, where he worked with malaria and black water fever, a dangerous complication to malaria that he put down to quinine poisoning.

In early summer 1898, Koch was back in Berlin but he only stayed there for a few months. He wanted to continue his malaria research and in the autumn he spent some time on the Roman Campagna, one of the worst malaria districts in Italy. In springtime next year he was back in Italy, this time together with his wife and in the autumn they travelled to Java. Koch became more and more convinced that Laveran and Ross were right about mosquitoes being the vectors for malaria. In German New Guinea, his wife became ill with malaria and had to go back to Germany since she was hypersensitive to quinine and could therefore not be treated for the disease. Koch and his collaborators continued their untiring struggle to eradicate malaria or at least to limit its ravages. It seemed hopeless to fight the mosquitoes in New Guinea, and Koch therefore decided to use another method.

The malaria-exposed population was examined, and those who had parasites in their blood were treated with quinine until the blood was free of the malaria plasmodium and they had no more relapses. This method, that required a good medical organization and disciplined patients, was introduced in all German colonies and in many instances proved to be effective. When Koch returned to Berlin in October 1900 in order to devote some time to his institute, he had spent three of the last four years travelling in the Tropics.

The next few years, Koch was engaged in combatting epidemics of typhoid fever in Prussia and the Ruhr using in principle the same methods that had been successful against cholera — bacteriological diagnosis, water hygiene and isolation of patients and symptomless carriers of the disease. When the epidemics petered out, Koch was again seized by the longing to visit foreign, exotic research areas that seemed constantly to lie in wait for him. It is very typical that when in 1904 he at last was elected to the German Academy of Sciences, he could not deliver the obligatory entrance speech until five years later. His excuse was simply that his many scientific expeditions had made it impossible to give the address at an earlier time. At the beginning of 1903, Koch and his collaborators, as always accompanied by his wife, travelled to Rhodesia to investigate yet another cattle disease, this time a tick-mediated piroplasmosis that Koch called "African coastal fever" and tried to immunize the animals against by repeated injections of blood containing the parasite.

In 1905, Koch was back in Dar-es-Salaam in order to study the life cycle of the piroplasma that caused the African coastal fever. He also worked with the trypanosome of the sleeping sickness and tried to elucidate its life cycle in the vector, the tsetse fly. A year later, Koch headed the German sleeping sickness commission in East Africa. His wife accompanied him again, but when they encamped in straw huts on the northwest shore of Lake Victoria, where sleeping sickness raged worse than ever, Koch resolutely packed his young wife off to Germany — this was something that he definitely refused to expose her to. In addition to its scientific work, the commission also attempted to treat patients with atoxyl, an arsenic containing synthetic substance that unfortunately in spite of its name turned out to be far from atoxic. Sometimes it caused atrophy of the optic nerve, and of 1633 patients treated with atoxyl, 23 became permanently blind. On the other hand, it was the most effective medicine against the deadly sickness available. Koch also seems to have contemplated some drastic preventive measures, for instance the extermination of all big game in the vicinity of large cattle herds. The wild animals were obviously hosts for the tsetse fly, but Koch's project caused an outcry of indignation among big game hunters and had to be abandoned. A somewhat unexpected finding was that the tsetse fly was very fond of sucking blood from crocodiles. Consequently, Koch devoted much time to avidly hunting crocodiles, a peculiar occupation for a bacteriologist.

Honors

Few medical scientists have been so honored as Koch and few have endured so much criticism. The list of his orders and other honors is impressive: The Grand Cross of the Order of the Red Eagle; *Pour le Merite* (the highest distinction for military valour in Prussia, which could also be awarded for more peaceful merits); membership of the Kaiser Wilhelm Academy with the rank of major general for his achievements as military hygienist; foreign member of the French Academy of Sciences, where he succeeded his old antagonist Rudolf Virchow; and last but not least, *Wirklicher Geheimrat* with the title *Excellenz*. Nowadays, it is of course the Nobel Prize in Physiology or Medicine of 1905 that must be esteemed the highest honor. However, the prize did not at this time have the enormous prestige that it enjoys today. In a later chapter, we will return to his Nobel Prize.

Koch would seem to have had all the praise that any scientist could ask for, but there were some jarring sounds among the ovations. It was not until 1904 that the German Academy of Sciences could make up its mind to elect him as a member. At that time, he had long enjoyed a standing in medical science that can only be called unique. Nevertheless, the Academy had dragged its feet and hesitated for decades, and when at long last it made up its mind, something astounding and rather embarrassing happened. The ballot was with black and white balls, like in a superior English club, and it turned out that of the 44 members present, no less than eight put a black ball in the ballot box. They did not consider Koch worthy to become "ordentliches Mitglied" (ordinary member) of the illustrious assembly.

Who were the members that had thrown the black balls and what were their motives? The ballot was of course secret and we will never know the answer, but it is tempting to speculate, at any rate about their motives. An obvious reason for the black-balling was of course the tuberculin scandal. Not only had Koch been wrong and inspired many patients and their doctors with false hopes, with his usual stubbornness he had stuck to his mistake as long as possible. In fact, Koch never unequivocally retracted his claims about the therapeutic effects of tuberculin. He just let it sink into what he probably hoped would be merciful oblivion. In the tuberculin case, Koch undoubtedly appears with all the human weaknesses that this admirable man unquestionably had. It is understandable if some of the Academy members could not forgive him that. At the same time, one can suspect other, not so honorable motives among the men of the black balls.

Koch had a sharp tongue, in a debate he was an effective, not to say caustic opponent. Moreover, in many cases he had proved himself right, an unpardonable fault in the opinion of many of his adversaries. One might have thought that they should nevertheless have been magnanimous enough to acknowledge his exceptional scientific standing. On the other hand, even his most ardent admirer must admit that Koch himself was hardly inclined to be noble and forgiving when dealing with his antagonists. Perhaps the juicy scandal of his divorce and remarriage to a much younger artist, had something to do with it? The social standing of a spouse (or lack of such standing) was rather important during this period. Even so, the readiness with which his young wife had subjected herself to dangers and hardships in order to accompany her husband under the most trying circumstances, should have made her more acceptable in the eyes of the Academy members.

The Man

Koch provoked strong feelings in his fellow beings; feelings of admiration and respect, in many cases devotion and friendship, but also envy and dislike, sometimes even hate. He was not really a modest person, nor was there any reason why he should be. However, he could sometimes give vent to an unexpected humility. In a speech in New York 1908, he explained his successes by saying that during his wanderings in the fields of medicine he had occasionally come to regions where gold might still be found at the wayside. Otherwise, it was perhaps self-confidence and perseverance that were his most characteristic qualities as a scientist. His motto, that he constantly impressed on his pupils, was *nicht locker lassen* ("to be dogged and never give up"). Sometimes the will never to yield could go too far. Tuberculin is the prime example here, but there are others. To the very last, Koch maintained that bovine tuberculosis constituted no threat to humans, in spite of the fact that both Emil von Behring and also British scientists had presented convincing evidence for a different opinion.

One could perhaps say that Koch was better at handling bacteria than human beings. Certainly, he might appear hard and insensitive in his dealings with colleagues who had failed to appreciate his achievements. At the same time, he had a very different side to his personality, which he demonstrated in his relations to old friends like Paul Ehrlich, and above all to his daughter Gertrud or "Trudchen" (a diminutive of her name) as he calls her in his letters. One of those letters was written in the Nile delta in November 1883, when he was waiting for the steamship that by way of the Suez canal would take him to India where the cholera still raged and offered ample opportunities for research. It is a long and rather touching letter in which the great bacteriologist apologizes for not being able to send her more desiccated flowers of the local flora, but the climate is so humid and the flowers do not keep. To comfort his young daughter, he has fixed a genuine mosquitoe to the letter and he gives detailed instructions how she should go about detaching it. Her dad especially points out that it is a mosquitoe of the very worst kind, the sort that you often got bitten by. Koch cannot have been entirely lacking in the psychology of everyday life, and one should perhaps also keep in mind that he was very highly regarded as a medical practitioner in Wollstein.

He felt a paternal responsibility for everyone working at the Institute for Infectious Diseases. In July 1903, he was staying in Bulawayo in Rhodesia when a telegram reached him with the news that a physician in Berlin, working with the plague bacillus, had been infected and died. Koch of course assumed that it must have been one of his collaborators at the institute and he was deeply worried. It was a great relief to him when he eventually learned that the victim was an unknown Austrian doctor who was in no way connected with Koch's institute. However, three years later, while Koch was encamped on the shores of Lake Victoria in East Africa, a very serious accident occurred in his own laboratory in Berlin while the staff was working with the trypanosome of sleeping sickness. Large quantities of trypanosomes were prepared by infecting rats and bleeding them when the sickness had culminated. The laboratory assistent Schmidt, who was supposed to hold the rat during the operation, injured himself while manipulating the animal and unluckily got his wound infected with trypanosome-containing blood. Treatment with atoxyl started immediately and Koch was notified. It is obvious from several letters which Koch wrote from Africa how deeply he engaged himself in the case and he also gave instructions about the treatment. The doctor in charge in Berlin, Bernhard Möllers, has written the most extensive biography of Koch, and in his book he is happy to relate that Schmidt completely recovered after the accident and still held the same position at the institute, when Möllers 35 years later finished his biography.

There were undoubtedly traits of human warmth and consideration in Koch's somewhat rugged personality. Surely, this was one of the reasons why he was able to build up a successful, internationally leading group, where many of the foremost bacteriologists of that epoch were trained. In Möllers' biography, there is a list of Koch's collaborators during these years that contains no less than 57 names. Many of these are outstanding figures in the history of bacteriology. This long line of pupils must be considered as proof of human qualities of a rather remarkable kind.

At the same time, it was very much a question of personality if one managed to be on a friendly footing with Koch. Among his early collaborators, there were quite a few who had a close and friendly relationship to Koch until the end of his life. The best known example is Paul Ehrlich. On the other hand, there is Emil von Behring whose originally close relation to Koch soon deteriorated and, when their research projects came on a collision course with each other, turned into real hostility. For collaborators whose personalities did not admit of real friendship with the great master, there were other and very strong motives for holding out. Certainly, some of his pupils found Koch an exacting master to work under, but after Pasteur's death he was without doubt the leading figure in bacteriology, the most important and the most rapidly expanding field of research in medicine. One could put up with a lot in order to be a member of such a research group as his.

In our opinion it is obvious that Koch's greatest contribution is the basic technique he developed that made the amazing progress of bacteriology in the last decades of the 19th century possible. There is no one who can compete with him here, he is the incomparable pioneer. However, it is possible that Koch himself did not agree with this evaluation of his achievements. Like his great rival Pasteur, he seems to have been motivated by a wish to see his research result in tangible progress in therapeutic and preventive medicine that immediately benefitted humanity. Occasionally, this ambition prevailed over his better judgement, as in the case of the tuberculin, but one cannot doubt his compassion on the victims of infectious diseases. He applied all his energy to combatting these diseases, which dominated the medical panorama at the time, not only in India and other exotic countries, but also in Europe. Koch undertook extensive scientific expeditions to the most distant and difficult parts of the earth, endured hardships and took risks in his work with tropical diseases that are quite astounding. His enemies, and he had many, claimed that he was just trying to escape from matrimonial scandals and all the trouble about tuberculin, but that is hardly the whole explanation. It might certainly have been a relief to take his young wife with him on journeys to the other side of the globe, far from the malicious slander in the drawing rooms of Berlin, but that could have been done under far more agreeable circumstances than among tsetse flies in Uganda or malaria-carrying mosquitoes in New Guinea.

One might have thought that Koch's restless activity would have required an iron constitution, but the fact is that he often had to take sick leave and make long recreation trips. It appears that Koch during his long stays in malaria-infected districts must himself have been stricken with the illness. In a letter of 1902, he writes about an attack of malaria that had required treatment with quinine. During his last years, his heart troubled him. He complained of being short of breath and found it difficult to walk on the stairs. In April 1910 he gave a lecture on the epidemiology of tuberculosis in the German Academy of Sciences. A few days later, he had a massive myocardial infarction from which he never really recovered. During a stay at the famous health resort in Baden-Baden, he died on 17 May the same year. He was found peacefully sitting in his easy chair, which he had pulled up to the open balcony door where he could enjoy the fragrance of the early summer flowers and look out over the Schwarzwald and the distant mountains.

EMIL VON BEHRING

The Discerning Vicar

The Prussia of the Hohenzollerns and its successor, the German Empire proclaimed in Versaille 1871 after the crushing victory over France, is often associated with militarism and an aggressive foreign policy, but it also had other and very different traditions. For instance, it is astounding and really quite impressive that the school for training of Prussian military surgeons, the Friedrich Wilhelm Institute, under a strongly science oriented leadership, became an important nursery for German biomedical research, and we have already seen examples of this with names like Rudolf Virchow and Hermann von Helmholtz. To this roll should be added also the name of Emil von Behring, the first in the long line of medical Nobel laureates.

Behring was born on 11 March 1854 into a large family where the father was an impecunious school master in a small village in what was then West Prussia. He had 12 children to provide for (one had died at the age of four) and the environment in which Emil grew up was characterized not only by the traditional Prussian piety and patriarchal sense of duty, but also by an exceedingly strained family economy that often must have bordered on poverty. The vicar of the parish, the Reverend Leipolz, noticed the boy's talents and saw to it that he was sent to the Gymnasium in Hohenstein in East Prussia. On one occasion his parents, because of their extremely poor economy, had decided to let Emil leave school, but his teachers persuaded them to change their decision by stressing the boy's intelligence and not least by promising economic help to meet the cost of his schooling.

Already in the Gymnasium, Behring had become interested in medicine, but his parents considered medical school as out of the question on economical grounds and instead wanted him to become a priest. He was about to set out for Königsberg to begin his theological studies when fate, once again in the shape of the vicar, for the second time intervened in his life. The Reverend Leipolz had a nephew, Oberstabsarzt[1] Dr. Blumensaht, who had come to visit his uncle, and on this occasion the vicar brought up the subject of Emil Behring and his wish to study medicine. Apparently, the Reverend Leipolz managed to persuade Dr. Blumensaht that here was a brilliant mind that must be saved for medicine. Dr. Blumensaht obviously had the right connections, and thanks to his efforts Behring was accepted as a medical student at the Friedrich Wilhelm Institute in 1874. Immunology has every reason to count both the Reverend Leipolz and

---

[1] A senior rank of military surgeon.

Emil von Behring (1854–1917).
Courtesy of the Nobel Foundation, Stockholm, Sweden.

Dr. Blumensaht among its benefactors when they saw to it that Emil von Behring did not end up as a parson in the German countryside.

A Military Surgeon

Training as a military surgeon at the Friedrich Wilhelm Institute meant a tough working program, but it was intellectually stimulating and offered opportunities for contact with medical research. The institute had close connections with the universities and young military surgeons could for years be allowed to work as assistants at well-known research laboratories and clinics. Behring seems to have enjoyed his time at the institute and after graduation he was appointed as assistant surgeon at a cavalry regiment in Posen. This might seem like a poor starting point for a scientific career, but during his ten years of obligatory service in the army Behring could devote a certain amount of time to his own research projects.

Behring took his duties as a military surgeon very seriously, and it seemed natural for him to work with wound infections and their treatment. At this time, there was a growing insight that both contagious diseases and local infections were caused by microorganisms. It therefore became important to find disinfectants to sanitize latrines and cesspools, but also for local treatment of infected wounds. Behring was fascinated by the possibility of finding disinfectants that could be used to treat not only wounds, but also infectious diseases. That of course required that the substance in question must be relatively harmless for the human organism when given by mouth or as injection. This was far from being the case with such classic disinfectants as for instance carbolic acid. He had begun to study iodoform, that had recently been introduced for the local treatment of wound infections, but he soon became convinced that it was far too toxic to be used for treating general infections like contagious diseases or blood poisoning.

These early, in themselves hardly important iodoform studies, at the same time show us how Behring's fundamental scientific thinking was taking shape. It is interesting to see how even at this early stage he seems to have developed the idea that iodoform does not kill the bacteria themselves, but instead in some still unknown way prevents them from forming poisons that might harm the organism. He becomes increasingly captivated by the thought that the organism itself must have the ability to destroy dangerous bacterial poisons in a way that leaves the body's own tissues and organs unaffected. These ideas of a specific defence mechanism would seem to portend the principles of the immune system.

Emil von Behring as a young man.
Source: Author's collection.

During his service as military surgeon, Behring in 1885 passed the examination for district medical officer and two years later he was appointed Stabsarzt,[2] a promotion that he must have been very proud of. It is typical of him and also a bit touching that in a letter of 1893, at a time when he was already famous in the medical world for his discovery of the antibodies, the letterhead designates him as "Stabsarzt, Professor Dr. Behring" — the rank of Stabsarzt comes first! In 1889 his superiors, who obviously had a high opinion of his scientific ability, assigned him as assistant to Robert Koch at the Department of Hygiene at the Berlin University. This gave Behring an opportunity to develop his talents in a unique scientific milieu and to take up the problems that he had already been thinking about but had not been able to tackle.

Antibodies

In the United States, the physiologist Henry Sewall had shown in the 1880s that pigeons could be made resistent to relatively large amounts of rattle snake poison if they had previously been treated with gradually increasing doses of the poison. The French bacteriologist Albert Calmette (1863–1933), best known for his vaccine against tuberculosis, could confirm Sewall's results and further showed that the blood of the comparatively resistant pigeons contained a factor that inactivated the snake poison. Two of Pasteur's most outstanding pupils, Émile Roux (1853–1933) and Alexandre Yersin had shown in 1888 that the diphteria bacillus, which had been isolated in 1884 by Koch's collaborator Friedrich Loeffler, formed a poisonous substance, a toxin, and that this toxin was responsible for the severe tissue damage that diphteria caused in the throat and larynx, and likewise after spreading by way of the blood, in the heart; damage that often led to the death of the patient. Soon afterward, Knud Faber in Denmark discovered that the tetanus bacterium, which can infect wounds particularly if they have been contaminated by earth, can also form a toxin that is disseminated in the blood and attacks the central nervous system.

Koch appears to have given Behring a fairly free hand to work with his ideas about a defence system of the body itself against bacteria and bacterial toxins. At this time, there were certain results obtained at the Department which indicated that blood could sometimes kill bacteria and inactivate bacterial toxins. Behring had for instance, together with some other researchers, found that serum from rat,

---

[2]Military surgeon with the rank of captain.

Shibasaburo Kitasato (1856–1931).
Source: Author's collection.

a laboratory animal that seemed to be resistant to infection with anthrax, could kill anthrax bacteria in the test tube. He considered that this ability was the reason why rats were resistent to anthrax. When this phenomenon was investigated more closely it was found to be species-specific. Sera from other species, that were not resistant to anthrax, did not kill anthrax bacteria in test tube experiments. Nor had rat serum any effect on a number of other bacteria that were tested. In other words, it was not a question of a general bactericidal ability common to all sera.

Working together with the Japanese bacteriologist Shibasaburo Kitasato (1856–1931), who since 1885 had been active at the Department, Behring in 1889 began to investigate the diphteria and tetanus bacteria and their toxins. He attempted to induce immunity in guinea pigs by infecting them with diphteria bacteria under conditions where a considerable part of the animals survived the infection. Behring found that the surviving guinea pigs had become immune, i.e. they remained healthy even after reinfection with diphteria bacteria using amounts that by experience he knew should have been lethal in animals that had not been immunized. Similar results were obtained in a series of experiments together with Kitasato where rabbits were infected with tetanus bacteria isolated by the Japanese scientist.

It turned out that the resistance against both diphteria and tetanus depended on something that was present in the serum of the immune animal. Perhaps the most interesting observation was that these immune-sera did not kill or inhibit the growth of the bacteria themselves. Instead they neutralized the bacterial toxins. These results seemed in an amazing way to verify Behring's old, intuitive ideas about a disinfectant, produced by the organism itself, that destroyed the toxins through which the bacteria caused disease. It is understandable that the role of bacterial toxins in causing infectious diseases was central for many years in Behring's thinking about how to prevent and cure such diseases.

Koch found Behring's and Kitasato's results interesting and important but nevertheless advised against publication before all the details had been carefully considered. It was not until December 1890 that Behring and Kitasato published an article in *Deutschen Medizinischen Wochenschrift* with the title *On the Origin of Immunity Against Diphteria and Tetanus in Animals*, a paper that in spite of its title mainly dealt with immunity against tetanus. Shortly afterwards, there appeared in the same journal another paper with Behring as the only author and the title: *Investigations of the Origin of Immunity against Diphteria in Animals*. With these two papers, an entirely new principle in immunology had been established, firmly based on the soluble components of the blood serum. As Behring himself often put it, one had gone from Virchow's, until then completely dominating

Source: Author's collection.

cellular pathology, to a resurrection of the old Hippocratic ideas. The principle of cellular immunity, that Elie Metchnikoff had advanced in the 1880s, now clashed with the new humoral immunity in the form of free, circulating antibodies. This new concept of immunity had an almost sensational breakthrough, not least because it offered the possibility of a therapy against some dreaded diseases.

Another of Koch's collaborators, Paul Ehrlich, discovered already in 1891 that also plant poisons like ricin and abrin could give rise to antibodies in mice, thereby making the animals resistant to these poisons. He used an immunization technique where the mice were given increasing amounts of the poison in a series of injections. Ehrlich coined the term active immunization for this process, that he considered analogous to Behring's and Kitasato's immunization against diphteria and tetanus toxins. The type of immunity obtained by giving the animal antibody-containing serum, he called passive. Ehrlich's results of course strongly supported Behring's discovery of the antibodies and the two scientists remained friends to the end. Ehrlich also worked out methods to determine the antibody level in sera so that they could be standardized, a crucial contribution to the new serum therapy.

The Serum Therapy

Behring had early on been conscious of the therapeutic possibilities that his discovery of the antibodies against bacterial toxins offered, what was now being called antitoxins. It ought to be possible to use immune-sera to obtain a passive immunization of the patients and prevent the devastating effects of the diphteria and tetanus toxins. Diphteria in particular was a dreaded disease with thousands of cases every year only in Berlin, most of them children, and a mortality of 30 to 50%. Thus, the need for the new therapy was not in question, the problem was how to make the immune-serum available to all these patients. The original experimental work on diphteria had been done with guinea pigs and much larger animals in great numbers were necessary in order to produce the amount of serum required for clinical tests. The supply of suitable animals was completely inadequate at the Department of Hygiene and all that Koch could offer Behring and his loyal collaborator, Stabsarzt Fritz Wernicke, was a single sheep. Little by little, things improved but the supply of animals suitable for immunization long remained a very difficult bottleneck. A great step forward was taken when Roux at the Pasteur Institute began to use horses for immunization on a large scale to obtain serum against diphteria toxin, a method that Behring immediately adopted.

Friedrich Althoff.
Source:Author's collection.

All great therapeutic breakthroughs have given rise to legends and that is the case also with the diphteria serum. Legend has it that on Christmas Eve 1891, a little girl lay dying of diphteria at the clinic of the famous surgeon, professor Ernst von Bergmann in Berlin. The reason why she lay at the surgical clinic was because the only treatment that could prevent the child from being suffocated by the diphteria membranes in the larynx was a surgical operation that opened the trachea (tracheotomy). The girl was not expected to survive the night and the nurse in charge recalled that professor von Bergmann had mentioned a certain Stabsarzt who should be sent for in desperate cases like this. A messenger was dispatched to the address given, the barracks of the Second Guards Regiment, and after a while the messenger returned accompanied by a tall figure in uniform. The unknown Stabsarzt took out a test tube containing a yellow liquid, which he drew up in a syringe and injected into the dying girl, whereupon he disappeared into the night. To everyone's astonishment, the little girl appeared healthy and well on Christmas morning.

Unfortunately this touching story, fit for any Christmas magazine, is probably pure fantasy. Even if Fritz Wernicke is an obvious candidate for the role as the unknown Stabsarzt, there was not, according to Behring himself, enough diphteria serum available to attempt any clinical treatments at this time. It was not until the beginning of 1893 that this became possible. The difficulty with the production of serum was first and foremost that Behring could not possibly keep a sufficient number of big animals for large scale immunization. The cost of feeding and housing them could simply not be accommodated within the budget of the Department of Hygiene. The fact that Behring already in 1892 had come in contact with a very influential high official in the Prussian Department of Education, Friedrich Althoff, who had been very favorably impressed by the young scientist and continued to support him in the future, was of no avail to solve this problem. The Prussian government was not prepared to assume the necessary expenses.

Both Koch and Althoff suggested that Behring should instead contact the German pharmaceutical industry and after some misgivings, at this time he apparently did not approve of taking out patents for medicines (he would later change his mind in this respect), Behring in 1892 made a contract with the company Höchst in Frankfurt am Main. He now had the economic prerequisites for serum production on a large scale, particularly after they had gone over to using the isolated toxin to immunize horses as suggested by Roux. The other great difficulty was to obtain sera with a high antibody level that could be analyzed by reliable methods. Even this thorny problem was eventually solved, not least through a close collaboration with Paul Ehrlich, who had become director

Emile Roux (1853–1933).

Source: Author's collection.

of a government institute for control of the serum production, which had been established in 1895 on the initiative of Althoff.

Clinical tests with diphteria antitoxin started in 1893 at several children's clinics in Germany that Behring supplied with sera. To begin with, the results were not entirely convincing, mainly because the level of antitoxin in the sera employed was too low and in addition varied in an uncontrollable way. When the quality of the sera improved, so did the results of the treatment. To give an example, Koch's collaborator Hermann Kossel was in charge of the ward that was attached to the Institute for Infectious Diseases recently created for Robert Koch. Here, Kossel made a series of careful clinical studies. When he was supplied with the new, highly active sera produced at Höchst, where the antitoxin level was a hundred times higher than in Behring's original preparations, his results were sensationally good. In May 1894, he reported on the treatment of 223 children with diphteria. When the treatment started within two days after the parents had noticed the first symptoms of the disease, 97% of the children survived. If the serum therapy started at a later time, the result of the treatment deteriorated and when the delay was as much as five to six days after the child was taken ill, the mortality was approximately the same as in cases without serum therapy. The number of tracheotomies was also much lower with early serum therapy.

The discovery of the antibodies and their clinical use was considered an enormous breakthrough by the entire medical world. French medical scientists headed by Émile Roux were particularly far advanced with regard to the production of diphteria antisera and their clinical use. So far advanced, in fact, that some patriotic French journalists got it into their heads that this was a purely French discovery. Behring was quite upset. He was very sensitive about what he considered to be encroachments and unwarranted claims to share the credit of his discoveries. As a result, he was often involved in heated priority conflicts. In this particular case he need not have worried since Roux, a thoroughly honest person, immediately explained the situation and gave all the credit to Behring. That was not all, when Roux was decorated with the Knight's Cross of the Legion of Honour, he refused to accept this honor unless his German colleague also received the same order. Consequently, in 1895 Behring was made an *Officier de l'Ordre National de la Légion D'Honneur*, an honour that he valued very highly.

The importance of Behring's work was early recognized in France and his relations with the Pasteur Institute, which after Pasteur's death was directed by Émile Duclaux and since 1904 by Roux, were always friendly and intimate. This also included Elie Metchnikoff, who already in 1888 had left the politically troubled Russia and obtained a position at the Pasteur Institute where he remained until his death. One might have thought that Behring, who could be rather aggressive when it came to upholding his scientific opinions, should have found it difficult to appreciate Metchnikoff and his cellular immunity, a theory that emphasized entirely different sides of immunity than those that Behring had worked with. In reality, Metchnikoff and Behring became close friends and a fairly extensive correspondence exists between them. The letters bear witness not only to scientific but also to warm human relations, which seem to have outlasted even the convulsions of the First World War.

The production and clinical testing of highly active diphteria antitoxin sera had claimed all Behring's time and energy so that he had not been able to take the same interest in the tetanus problem. Another reason may have been that it was the struggle against diphteria which had been admired all over the world. Diphteria was a clinical problem of an entirely different order of magnitude than tetanus, and in addition it was an illness that primarily affected children. For many years, letters continued to reach Behring from grateful parents who in the most touching way expressed their gratitude for what he had done to save their grievously sick children. This must have been a source of the greatest satisfaction to Behring. In addition, he undoubtedly regarded the diphteria antitoxin as entirely his own contribution, which might have made him inclined to give it priority. The credit for the discovery of the tetanus antitoxin he had to share with Kitasato, even if the latter had already returned to Japan in 1892.

In times of peace, tetanus was a rare disease compared to diphteria. On the other hand, the mortality was around 90% and in wartime, when deep wounds easily became contaminated with earth, the number of tetanus cases increased dramatically. However, when Behring and Kitasato made their discovery, Germany had experienced a long period of peace and consequently the demand for tetanus antitoxin was rather low. To begin with, Behring had thought that it would be possible to treat an already manifest tetanus with antiserum, the way one could do with diphteria in children. This turned out to be impossible with tetanus. Not even high doses of antiserum had any effect once the symptoms of tetanus had set in. Since this typically does not occur until between one and two weeks after the injury, it should have been an obvious idea to give tetanus antitoxin during this free period in order to prevent the outbreak of the illness. Strangely enough, it took a long time before such prophylactic treatment was used routinely, and it was in fact the veterinarians who took the lead here. Even so, it was not until 1895 that a veterinarian reported the favorable results of a prophylactic treatment with antitoxin of horses, the domestic animal most liable to tetanus infections.

In spite of its use in veterinary medicine, the demand for tetanus antitoxin continued to be low and the production of antiserum took place in Behring's own laboratory in Marburg, where he was a professor of hygiene since 1895. The Höchst Company, that had produced great amounts of diphteria antiserum, was never interested in tetanus. It would not be until the outbreak of the First World War that there was any real interest in Behring's tetanus antitoxin. In the beginning of the war, there was a serious shortage of tetanus antiserum in the German army and it has been estimated that something like 0.5% of the wounded got tetanus. A report from a field hospital in Namur shows that during the period 11 September– 30 November 1914, of 2193 wounded brought in 27 (1.2%) got tetanus. After obligatory serum prophylaxis had been introduced, there was not a single case of tetanus among 1195 wounded. There is no doubt that Behring's antiserum, which was used on both sides during the war, must have saved the life of countless wounded soldiers. Not without reason, Behring on 15 October 1915 was awarded the Iron Cross, second class to be worn in a black and white band. This was undoubtedly a well deserved honor if one considers it meritorious to save the life of soldiers, not just to kill them. Even if heaps of the Iron Cross were awarded in the German armed forces during the First World War, it was unusual that it went to a civilian and Behring seems to have been glad of the decoration.

Active Immunization Against Diphteria Toxin

Behring's discovery of the antibodies and their use in the clinic had transformed diphteria from a mortal threat, that was a terror to all parents and claimed countless victims among children the world over, to a serious but treatable illness with a fairly good prognosis if serum treatment was given early enough. However, the use of the diphteria antitoxin had no effect on the prevalence of diphteria in the population (the morbidity). It was a question of passive immunization against the toxin which did not begin until after the outbreak of the illness. Attempts to prevent the spreading of the disease by isolation of the patients, closing of schools and disinfection of places where diphteria cases had occurred were of no avail. Instead, the morbidity continued to increase. In 1906, there were 2997 cases in Berlin, but already in 1911 the number had risen to 11,578.

Behring was of course well aware that the morbidity of diphteria continued to increase, even if the prognosis was now incomparably better. He had early on entertained the idea of trying to make an attenuated toxin that could be used for the immunization of humans. Extensive experiments had been made in 1894 but failed to obtain a satisfactory vaccine. He also tested treatment of the toxin with formalin, a method that much later would be used to inactivate the toxin without interfering with its immunization ability. However, Behring was not satisfied with the results. It appears that he simply could not bring himself to believe that a completely inactivated toxin could be used for immunization. When in the years just before the First World War, he again took up the problem of active immunization against diphteria toxin, he therefore used another method that had been employed in the immunization of horses for serum production.

The principle was to use a mixture of toxin and antitoxin where most of the toxin had been neutralized, but with enough active toxin remaining to give immunity. The vaccine had been stabilized by the addition of formalin, which we now suspect must have been of great importance in making it harmless and at the same time efficient. Behring presented his vaccine at a congress for internal medicine in Wiesbaden in 1913 and already the next year there were encouraging results. Of 633 children vaccinated, only two became ill during a number of local epidemics, while among the unvaccinated children in the neighborhood, there were a considerable number of cases, as one might have expected. The active immunization against the toxin seemed to have prevented an outbreak of the disease among the vaccinated children also in such cases where diphteria bacilli could be demonstrated in the throat. The vaccination had no effect on the presence of bacilli in these children who continued to have positive tests for diphteria.

After the outbreak of the war Behring's laboratory in Marburg had to concentrate on the production of tetanus antitoxin and the work on the diphteria vaccine had to take second place. However, the military authorities eventually became aware of the fact that epidemics of diphteria were a considerable problem also in the armed forces and consequently there was a renewed interest in the vaccine. The active immunization against the diphteria toxin was the last of Behring's great scientific contributions before his health, that steadily deteriorated during the war, made further work impossible.

Honors and Conflicts

Behring undoubtedly had a pugnacious nature and never shunned a conflict however famous and influential his adversary was. It did not take long before his originally good relations with Robert Koch had deteriorated, and by 1894 it was already quite tense. This may well have contributed to Behring's desire for a chair at some university instead of the position that he had at Koch's Institute for Infectious Diseases. As always, he was strongly supported by Friedrich Althoff, who made several attempts to obtain a suitable professorship in hygiene for his protegé. The problem was that the faculties he contacted were not entirely enthusiastic. To a certain extent, this can be explained as academic conservatism and a generally negative attitude towards the new subject, bacteriology. They simply wanted a representative of classic hygiene, which often enough was hostile to Koch and his school. Another objection to Behring was that his lectures were on a too high scientific level and unsuitable for medical students. Perhaps Behring's reputation as an aggressive and difficult person also had something to do with their attitude.

However, Althoff was not so easily put off. He was a powerful man and was used to having his own way. Behring was already a titulary professor — Althoff had seen to that — but now it was a question of getting him a chair at a university. A professorship in hygiene became vacant in Marburg and in spite of some opposition in the faculty, Althoff succeeded in having Behring appointed to the vacant chair in 1895. The new professor seems to have been happy with his position in Marburg, in any case he stayed there until the end of his life. A year after his appointment, he married Else Spinola, the 18-year-old daughter of the director of the large Charité Hospital in Berlin. Her husband may have been a little late in his début as a family man (the age difference between husband and wife was 24 years), but once he got started on raising a family the number of children increased rapidly, and in the end he had six sons. At the christening of the eldest son, Émile Roux served as his godfather, while in the case of the fifth son, Elie Metchnikoff performed the same function. A good example of the close relations between Behring and the leading men at the Pasteur Institute.

Not only did Behring's family life prosper during the years around the turn of the century, he was also awarded a number of honors. He became *Wirklicher Geheimrat* with the title *Excellenz*, and in 1901 he was knighted by the Kaiser, who was one of his admirers. Later the same year, he was awarded the first Nobel Prize in Physiology or Medicine, in our eyes the greatest accolade of them all. At the same time, it was during these years of triumphs that he became involved in a number of conflicts which all had a common denominator — his restless search for a specific treatment of tuberculosis.

In the beginning of the 1890s, Robert Koch was the generally acknowledged leader of research on tuberculosis. His tuberculin seemed to present an effective remedy against early stages of tuberculosis and in the beginning everyone was enthusiastic, with the exception of course of his old adversary Virchow. However, the critical voices gradually grew louder and in a couple of years they reached gale force. In 1895, Behring appeared before a congress of physicians and natural scientists in Lübeck and passionately defended his old teacher. At the same time, he launched the thesis that tuberculin functioned by causing the formation of an antitoxin in the same manner as diphteria and tetanus toxins did. The way tuberculosis and perhaps also other severe infectious diseases worked was the same as for diphteria and the principles of a successful treatment ought to be similar. There was a particular tuberculosis toxin that must be isolated, so that an antitoxin against it could be produced. A daring theory without any real experimental support, one might think, but also a line of thought that must have come naturally to Behring. He had, after all, recently opened up a completely new field in immunology and of course it was tempting to generalize his great discovery.

For many years, Behring looked in vain for a tuberculosis toxin which might be used for immunization and the production of an antitoxin practicable for the treatment of the disease. Several times he believed himself close to the goal, but in the end he was forced to admit defeat. In a publication where he summarized the results of the years 1895–1900, he resignedly concluded that all hopes of a therapeutically useful antitoxin had to be abandoned. This does not mean that the tuberculosis research that Behring continued until his long sickness period 1907–1910, should have been without results, even if the main goal could not be achieved. Unfortunately, there were also unwanted consequences in the form of scientific conflicts. The best known is his bitter conflict with Robert Koch.

In his search for a tuberculosis toxin, which might be used for immunization, Behring and his collaborators came up with certain preparations which in Koch's opinion were very close to his own tuberculin. In 1897, he therefore submitted an application for a patent on tuberculin, which made Höchst Company, closely related to Behring, contest Koch's application. The long patent litigation that followed resulted in a complete defeat for Koch. His relations with Behring were not inproved when their interests collided in veterinary medicine, where they both tried to find a way to immunize cattle against tuberculosis. Had this succeeded it would undoubtedly have been of great economic importance for cattlebreeding, but above all they hoped for new methods to treat and prevent tuberculosis in humans.

We have seen how Behring at the turn of the century had been forced to realize that the idea of a tuberculosis toxin had to be abandoned. Instead, he tried to obtain preparations from the tubercle bacillus that could be used to immunize primarily cattle against the bacillus itself, thus making them resistant to tuberculosis. Behring saw this as corresponding to Jenner's vaccination against smallpox and he in fact used the term "jennerization" to designate it. The intense efforts in Behring's laboratory resulted in no less than nine different preparations from tubercle bacilli intended to be tested as vaccines. This meant of course that they all had to be tested systematically on livestock, which presented considerable practical difficulties. The indignant Behring was convinced that Koch, through his minions among the veterinarians, tried to use his influence to obstruct Behring's research. The results of Behring's efforts to immunize cattle against tuberculosis were already judged rather differently at this time. Perhaps, one might summarize his tuberculosis research by saying that its greatest importance was its spin-off effects.

There was a marked difference of opinion between Koch and Behring regarding cattle tuberculosis. Koch had originally believed that both bovine and human tuberculosis were caused by the same bacillus, but he eventually realized that it was a question of two different bacilli. He also became convinced that for practical purposes, one need not consider the possibility that bovine tuberculosis could be transmitted to humans. Instead, he stressed the importance of human expectorations for the transmission of the infection. Behring, on the other hand, believed (mistakenly) that identical bacteria caused both human and bovine tuberculosis and he was convinced that the infection could be transmitted from the cow to human beings, and that cow's milk was one of the most important sources of infection as far as children were concerned. He had a point there, but he went one step further and maintained that human tuberculosis, even when it was first observed in adulthood and could not be demonstrated in the intestines, was originally a milk mediated infection that occurred already in infancy — a daring generalization to put it mildly.

Obviously, means must be found to sterilize cow's milk in order to eliminate this dangerous source of infection. A great problem for Behring was that he could not accept pasteurization (heating to a certain temperature) of the milk, since he was of the opinion that this procedure destroyed important immune bodies that protected the infant against intestinal infections. He instead tried a number of bactericidal substances, like for instance formalin, formic acid, and hydrogen peroxide, but none of these turned out to be practicable. Nevertheless, Behring made an important contribution to our knowledge of tuberculosis when he emphasized, against the authority of Koch, the importance of bovine tuberculosis and cow's milk as a source of tuberculosis in humans. If he never succeeded in his ambition to produce a useful vaccine, it was certainly not because of a lack of work and effort. Furthermore, it must be remembered that this is a problem that has still not been satisfactorily solved.

A Tormented Man

At the previous turn of the century, gentlemen who posed for photographers hardly tried to make a jolly and easy-going impression. The ideal of a man was far from ingratiating and smiling, rather it was stiff, imposing and dignified.

Nevertheless, there is something special about the photographs that exist of Behring even as a young man. He looks so doggedly determined, perhaps even tormented by his ambition, as if in spite of all his great successes, he never had experienced any real happiness in life. Looking at pictures of his great teacher and later enemy, Robert Koch, an entirely different personality seems to emerge. Certainly, he looks rather a gruff and no-nonsense character, but he also gives the impression of an inner strength and confidence in himself, of never having seriously doubted his own ability. He had indeed been involved in scientific conflicts, but unlike Behring he posessed a stable core of equanimity and self-reliance. The general impression is very far from that of an unhappy and tormented human being.

During the years 1891/92, Behring had periods of what he, in a letter to an old friend from his student days, describes as overwork, but where one has a feeling of a depression underlying his troubles. On such an occasion, he was sitting alone in a Berlin restaurant with a bottle of wine, in a mood that Winston Churchill, himself a victim of depressions, used to describe as "the black dog." A young assistant at the Institute for Infectious Diseases was rash enough to ask in a very polite way if he might be allowed to join the Stabsarzt at his table. Without even looking up, Behring answered with the single word: "No!" To the poor assistant this was obviously a shattering experience that he vividly remembered even at an advanced age.

In 1907, Behring wrote an article under his own name in *Berliner Tageblatt* where he repudiated anonymous allegations in the French newspaper *Figaro*, that because of a depression he had been admitted to a nursing home. Later it was even alleged that it had been a psychiatric clinic. Behring was very upset and saw it as an attack by his scientific adversaries. As a kind of unintentional confirmation of these malicious rumors, he in November of the same year consulted Dr. von Hoesslinsch in Munich, and was admitted to his clinic because of a deep depression complicted by severe pains in a foot. No somatic explanations of his pains could be found, nor was there any improvement of his mental condition during his stay at the clinic. He several times made up his mind to break off the seemingly meaningless treatment, but eventually he nevertheless remained in Munich. It was not until 1910 that there was a slow improvement in his condition and in August he was able to return to Marburg.

In his attempt to come up with a practicable vaccine against tuberculosis, Behring had worked with his usual feverish energy and driven his collaborators to their utmost for years. His illness came at a time when it must have become increasingly clear to him that his great project could not be realized. He would never be the man who eradicated tuberculosis, the goal that he had dreamt of ever since the middle 1890s and that had led to the heart-rending conflict with Robert Koch. Behring had probably suffered much more from their enmity than the rather stable Koch. It seems reasonable to see a connection between his general situation in 1907 and the prolonged period of depression and incapacity for work that now followed.

Behring was a very complicated human being, who had great problems with himself and occasionally also with his colleagues and collaborators. Even so, he undoubtedly had the gift of friendship. He had a close relationship with Paul Ehrlich, from their first acquaintance in Koch's laboratory until Ehrlich's premature death, even if their friendship was sometimes marred by Behring's tendency to make too great demands on Ehrlich. This created problems and conflicts between them, particularly when it came to the extensive testing of Behring's many tentative tuberculosis vaccines. Nevertheless, their friendship endured and at Ehrlich's funeral in 1915 Behring, who was himself not in the best of health, made a deeply touching speech in which he emphasized not only the unique position of the deceased in medical science but also the deep personal role that he had played in Behring's own life. Another example of Behring's talent for friendship is of course his close ties to the Pasteur Institute, in particular to Émile Roux and Elie Metchnikoff. Undoubtedly, both Ehrlich's death and the difficulties of maintaining the close relations with his French friends, made the war years even harder for Behring. In addition, his health deteriorated drastically.

In August 1913, Behring had an accident which resulted in a fracture of his left thighbone. According to the surgical principles of the time, he was placed in traction and it took a month before he could leave the hospital and return home. The fracture proved to be considerably more serious than had originally been suspected. It not only did not heal, it resulted in a false joint that made him an invalid and also caused him great pain. In September 1916, an abscess made an operation necessary, after which Behring became bedridden and never recovered. Until his death on 31 March 1917, he was attended to by a young surgeon, Georg Magnus, who grew quite close to Behring during the last months of his life. To entertain his patient, Magnus played chess with him, and in spite of Behring's often severe pains he was always the superior chess player, calm and coldly calculating. Magnus was struck by the older man's amazingly wide reading and extensive education. They had long discourses about the most varied subjects, mostly literary and philosophical, but rarely medical. Occasionally, these conversations turned into rather animated discussions where Behring often gave vent to a pessimism, which Magnus to the best of his ability tried to argue against, most of the time without much success. He had the impression that Behring was far from being an admirer of the policy which had caused Germany to ally itself with Austria in the First World War. On the contrary, Behring seemed to be at heart convinced that the war would end badly for the Central Powers.

When Magnus with the best of intentions tried to cheer up his famous patient by talking with admiration about his great medical achievements, Behring curtly rebuffed him with the words that in one's old age (Behring was 63 years old at the time), a life's work, however outstanding, could not be quoted as a merit; it was already too far behind you. On one occasion, Magnus had felt himself unusually hard beset by Behring's pessimistic sarcasms and he tried to defend himself by using a well-known quotation from Faust implying that redemption is possible for anyone who *immer strebend sich bemüht* (approximately "always assiduously exerts himself"). Behring had remained silent and lost in thought for quite a while and then he said, plainly and unaffectedly as if he at long last had found justification for his life: "And that I suppose I have done."

PAUL EHRLICH

Great scientists are not always nice and appealing characters, rather they can sometimes be quite difficult, not to mention unpleasant to people around them. Robert Koch and Emil von Behring may be cited as examples here. It is somewhat ironic that their mutual friend, who was extremely close to both of them, should in every respect be so very unlike them. While they were pugnacious and aggressive, always ready to take on both their scientific adversaries and each other, Paul Ehrlich was considerate and passive, perhaps even submissive. If both Koch and Behring had something stern and remote about them, that could intimidate others and cause estrangement, Ehrlich was charming and humorous, a kindly person whom people became spontaneously fond of, with something akin to a child's disposition. At the same time, he is one of the truly greatest figures in biomedical research with remarkable breadth and imagination. In an obituary of 1915, Sir Robert Muir laconically summarizes his opinion, that most scientists would agree with: "Ehrlich must be with the greatest, however small that company may be."

The Man Who Loved Colors

On 14 March 1854, Paul Ehrlich was born in the little town of Strehlen not far from Breslau. His Jewish family, which had lived here since the 18th century, run a successful corn trade that Paul's grandfather had expanded to include a general store and a liqueur distillery. Paul grew up in economically favorable conditions with a kind-hearted and somewhat even-tempered father, Ismar, and an energetic, business-minded mother, Rosa, née Weigert, who dominated the family. Like his father, the young Paul was shy and compliant, a bookish child who preferred to spend his time reading in his grandfather's extensive library rather than participating in the wild games of his schoolmates, where often enough he became the object of scorn and ridicule. As might have been expected, he was a diligent schoolboy who was particularly good in mathematics and the natural sciences. His grandfather's distillery with all its chemical apparatus had obviously made an indelible impression on the boy and this experience probably laid the foundation of his lifelong interest in chemistry. On the other hand, he was not a consistently successful student in all subjects. Both at school and later on as an adult, he had difficulties expressing himself in German both orally and in writing. The same was true of English and French where his accent was highly original to say the least. This would cause some complications in years to come, when as a world-famous scientist he had to present his results before an international audience.

Paul Ehrlich (1854–1915).
Courtesy of the Nobel Foundation, Stockholm, Sweden.

Paul Ehrlich's student years coincided with the almost explosive development of organic chemistry during the latter half of the 19th century and the synthesis of a great number of dyes that laid the foundation of the German chemical industry. In addition to causing a revolution of women's wear, these synthetic dyes would also prove useful for staining cells and tissues, and this is where Ehrlich made his first important scientific contribution. Ever since his schooldays, he had been fascinated by dyes and the possibility of using them in medical research, and he was encouraged by his mother Rosa's cousin Carl Weigert (1845–1904), who was only nine years older than Paul but already on his way to become an outstanding pathologist. During the holidays, Paul conducted experiments in Strehlen with anilin dyes that he mixed into the feed for his mother's domestic white pigeons. The idea was that they should assume a nice blue colour, but the most obvious result of the experiment was that the pigeons died. The intended result, in this case the change of color, had an unwanted side effect. This early experience made a deep impression on the young Ehrlich and illustrated a central problem in chemotherapy, the medical field that he would devote a great part of his life to develop.

Like so many other young men, Paul Ehrlich had difficulties deciding what career he should choose. After graduating from the Gymnasium, he studied natural sciences somewhat unsystematically for a term at the University of Breslau. His family was concerned about his future. Why could he not become a doctor, that way he would at least have an economically secure future? Ehrlich had a strong inclination towards medical research, but the actual practice of medicine had no appeal for him. With his sensitive, not to say timid disposition, he was reluctant to witness the suffering of his patients, and perhaps also not willing to inflict pain on them during the treatment. In the end, coupled with the strong support of Carl Weigert, his family managed to persuade him. During his medical studies, first in Strasbourg where he passed a preparatory examination (*Physikum*) and after that in Breslau, he became more and more fascinated with dyes and their medical uses. Carl Weigert had already established himself in this field, where he became a pioneer in the staining of tissue sections for microscopic examination in pathology. Through him, Ehrlich came in contact with the botanist Ferdinand Cohn and the outstanding pathologist Julius Cohnheim, both of whom were active in Breslau. In Cohn's laboratory, Ehrlich met Robert Koch for the first time, when he arrived in a horse and carriage from his rural practice in Wollstein in order to demonstrate the results of his anthrax research.

Ehrlich passed the examinations for his MD in 1877 and the next year he completed a thesis in Breslau with the title *Some Contributions to the Theory and Praxis of Histological Staining*. In his thesis, he reported the discovery of an, until then, unknown kind of cell, which he called mastcells, that surrounded the blood vessels and contained granules that could be stained with basic dyes. The discovery of the mastcells is of course of considerable interest as such, but it also illustrates a line of thought which would become central in Ehrlich's research. He reasoned that the biological effect of a substance, for instance a dye, is dependent on a chemical affinity between the substance and different structures which occur in cells and tissues. This chemical affinity is specific and it must consequently be possible by systematic testing to select substances suitable as drugs, i.e. with well-defined biological effects and no unwanted side effects. For instance, among the seemingly endless number of new compounds produced in the organic chemical laboratories, there must be some that kill or inhibit certain bacteria, but are harmless to the human organism.

Ehrlich's discovery of the mastcells and his new methods for the staining of cells and tissues had gained him international recognition. The head of the Medical Clinic II at the well-known Charité hospital in Berlin, professor Friedrich von Frerichs, offered the young physician a position at the clinic with excellent opportunities for research. Frerichs gave his new *Oberarzt* complete freedom to choose his own line of research, in accordance with one of the professor's leading principles: "Science is a bird that only sings in freedom." Here, Ehrlich spent seven happy years while he worked on improving his staining methods and made fundamental contributions to hemathology. During this time, he married Hedwig Pinkus, the daughter of a wealthy textile mill owner. She was ten years younger than her husband and their apparently happy marriage was blessed with two daughters whom he was deeply attached to.

In 1885, Ehrlich published the results of an investigation of the uptake of oxygen in the animal organism. By staining living organs and tissues (vital staining), he could show that they had a markedly varied ability to take up oxygen, with the highest uptake in different kinds of nervous tissues. These results were received with great interest, and two years later he got a prestigious prize for his discoveries. Nevertheless, the year 1885 was an unlucky one for Paul Ehrlich. His professor and enthusiastic patron, Friedrich von Frerichs, died suddenly and the new head of department, Karl Gerhardt, was a clinician of a more conventional type. He was not at all prepared to give Ehrlich the rather priviledged position that he had enjoyed under Frerichs and that had been a prerequisite for his successful research. Ehrlich found his position at the clinic as well as his research opportunities increasingly unsatisfactory and this probably contributed to the deterioration of his health. He had a persistent cough, which was hardly surprising since Ehrlich always seemed to have a cigar in his mouth. In any case, he was concerned about his health, and in 1888 became convinced that he had detected tubercle bacilli in his expectorations. Ehrlich suddenly left his position at the Charité and with his young wife he set out for the dry desert climate of Egypt in an attempt to cure his supposed tuberculosis. A year later he returned to Berlin, apparently cured of tuberculosis but without a position which would allow him to continue his interrupted research.

The Serum Institute

The only remaining solution was to establish himself as a private researcher. Fortunately, Ehrlich's father-in-law generously provided the means for this. Not far from his private residence, he rented a couple of simple rooms where he established his laboratory. This was a new beginning for him as he also embarked on a completely new line of research. It was now that he started his investigations of the plant poisons, ricin and abrin, and their ability to cause the formation of antibodies in mice. He defined the concepts of active and passive immunization and showed that antibodies could be transmitted through the maternal milk. This research naturally led to a collaboration with Emil von Behring. A close friendship, that would last their whole lives, evolved between the two scientists, who were in every respect so very different from each other.

The methods that Ehrlich devised for the determination of the antibody level in serum, and for the production of highly active diphteria antitoxin were integral for the success of Behring's serum therapy against diphteria (see the section on Emil von Behring). When Ehrlich in 1897 published his methods he included, as a kind of appendix, a theory of how antibodies are formed, his famous side-chain theory. He assumed that cells with the capacity to form antibodies, have receptors on their surfaces in the form of side-chains, which can bind substances that are alien to the organism (antigens), should such substances come in contact with the cell surface. This binding presupposes that there is a chemical affinity between antigen and receptor, in other words it is an example of Ehrlich's fundamental idea about chemical specificity as the basis for all biological functions. When the receptor has bound the antigen, the cell is stimulated to form a great number of new receptors that are shed to the blood and appear as free, circulating antibodies specific for the antigen in question.

There was great interest in Ehrlich's side-chain theory, but it was also critisised as being too "chemical" and fanciful. This criticism, put forward among others by the famous Swedish chemist Svante Arrhenius, was aimed at Ehrlich's preference for advancing over-simplified chemical models where he, for instance, assumed that the interaction between toxin and antitoxin led to the formation of stable chemical bonds, while Arrhenius instead maintained that these associations were relatively unstable and easily dissociated. Undeniably, Ehrlich introduced a number of new concepts in immunology that might appear rather hypothetical. In spite of this (at least to some extent) well-founded criticism, there is no doubt that when we examine his entirely speculative side-chain theory today (which when it was proposed had no real experimental support), it nevertheless in many ways portends the fundamental concepts of our modern views of how antibodies are formed.

Ehrlich had been in contact with Robert Koch ever since the former had helped improve the method for staining the tubercle bacillus in 1882. It is also likely that he tested Koch's tuberculin on himself when he believed that he had been infected with tuberculosis. Undoubtedly, he had been convinced of the effectiveness of the treatment, and during a period in 1890/91, he worked himself in the Moabit Hospital where the tuberculin was being tested clinically. At this time, when severe criticism against the therapeutic use of tuberculin was being voiced, Ehrlich was one of its strongest defenders. In later years, he was more guarded and thought that Koch had been premature in his attempts to use tuberculin to treat the disease. Generally speaking, the early 1890s was a period of naive optimism and enthusiasm for both prophylactic and therapeutic progress in medicine. Thus, in 1892, the well-known German philosopher Eduard von Hartmann wrote in a letter to Emil von Behring: "What should mankind undertake when all diseases have been eradicated?" Hardly a problem that we would worry about today. However, already during the first decade of the new century it became increasingly obvious that both active immunization (vaccination) and Behring's serum therapy was far from being the magic bullet that had originally been believed. On the contrary, there were reports of anaphylactic reactions when the serum treatment had to be repeated.

When Koch established his Institute for Infectious Diseases in 1891, Ehrlich was given a laboratory at the institute, and here he worked without pay for three years. The fact that he had a professorship at the Berlin University did not help his economic situation since it was an unpaid position. We owe it to his generous father-in-law that Ehrlich could continue his research without any real financial difficulties during these years. In view of his steadily growing international reputation as a scientist, Ehrlich's position was nevertheless somewhat embarrassing to the Prussian government. At least that was the opinion of Friedrich Althoff, who in reality made the decisions about governmental support of medical research from his central position in the Prussian Department of Education. He had already helped Behring to a chair at Marburg and now he intervened to secure reasonable working conditions for Ehrlich.

On Althoff's initiative, a special unit for the control of antitoxin production was created at Koch's institute and Paul Ehrlich was put in charge of what was later named the Institute for Serum Research and Testing. It was soon moved to new premises in Steglitz, a suburb to Berlin, where it was housed in exceedingly simple, not to say primitive rooms. With his characteristic modesty, Ehrlich was nevertheless very satisfied with his new institute and much of his fundamental work on the determination of antibody activity as well as the elaboration of his theory about the principles of immunization was carried out here. Althoff, however, rightly felt that Ehrlich deserved both strong encouragement and a better equipped laboratory. In 1897, Althoff saw to it that his protégé was appointed *Geheimer Medizinalrat*, and after negotiations with the mayor of Frankfurt am Main, the Royal Prussian Institute for Experimental Therapy was established in 1899 with Paul Ehrlich as its director, a position that he held until his death. During his time as leader of the Serum Institute in Berlin, his collaboration with Behring had sometimes led to conflicts as Ehrlich thought, and not without reason, that Behring was much too demanding. However, their old friendship survived this strain. It was typical of the egotistical trait in Emil von Behring to demonstratively refuse to attend the formal opening of the new institute in Frankfurt. The reason was that its name suggested to Behring that Ehrlich was now trying to free himself from his close collaboration with him.

The Advent of Chemotherapy

The idea of a chemical explanation of both the normal functions of the human organism and its diseases, can be traced all the way back to Paracelsus (1493–1541), with his notion of the importance of the chemical "principles" sulfur, mercury and salt, and their mutual balance, for the health of the individual. He also assumed the existence of specific remedies (arcana) indicated by God himself through some characteristic, for instance color and form of a flower. Paracelsus also introduced the use of inorganic chemicals, such as heavy metals and their salts in medicine, of small benefit to his patients. Thus, the idea of examining the chemist's arsenal of different compounds in order to find suitable drugs is nothing new. What distinguishes Ehrlich from his predecessors and makes him so successful is the systematic and critical way he goes about his task.

The choice of Frankfurt am Main as the place for the new institute was certainly not a random one. Here, the laboratory was close to a flourishing chemical industry that would become an important foundation for Ehrlich's

Paracelsus (1493–1541).
Courtesy of Bonnierförlagen, Stockholm, Sweden.

research — which the industry also supported economically. Franziska Speyer, the widow of a wealthy banker, donated in 1906 a large sum of money to set up an institute for chemotherapy in memory of her late husband, Georg Speyer, with Paul Ehrlich as its director. At the opening ceremony of the so-called Georg-Speyer-Haus, Ehrlich gave a speech in which he promised that now the chemists could synthesize compounds that would have effect only on the parasites which attacked the sick organism. These synthetic remedies would work as magic bullets that automatically found their target without causing any harmful side effects. A promise that may seem rather typical of the scientific optimism of the period as well as its propensity for flowery eloquence. Nevertheless, his words would prove to be, to some extent, prophetic.

In 1904, Ehrlich and his Japanese collaborator Kiyoshi Shiga had found that mice infected with trypanosomes could successfully be treated with the dye trypan red. In cases where the trypanosomes became resistant to trypan red, they proved to be sensitive to atoxyl, an arsenic-containing compound that Koch used to treat sleeping sickness during his African expeditions. The problem was that atoxyl sometimes caused blindness. When Ehrlich and his collaborator Alfred Bertheim had succeeded in determining the correct structure of the compound (p-aminophenylarsonic acid) they started to systematically synthesize hundreds of derivatives of atoxyl in the hope of increasing its effectiveness against trypanosomes and reducing its toxicity. Eventually these efforts, so typical of Ehrlich's methodical way of looking for therapeutically useful substances, resulted in the first great breakthrough in the treatment of the dreaded venereal disease syphilis.

Ever since the days of Fracastoro, syphilis had without much success been treated with preparations containing mercury, treatments that often enough caused more suffering to the patients than the illness itself. Even if the treatment was little better than during the renaissance, the early 20th century would see decisive progress in terms of our understanding of the cause of syphilis. The German zoologist Fritz Schaudinn (1871–1906) had specialized in the study of protozoa and their importance as parasites which caused human illnesses. This made him work for better contacts between medicine and protozoology, and when in 1901 he became director of a German-Austrian laboratory on the Dalmatian coast, he did important research on both malaria and tropical amoeba dysentery. Schaudinn was appointed director of the Institute for Protozoology at the German Imperial Ministry of Health in 1904, and here he started to work on syphilis. The reason was that he had been intrigued by earlier claims in the literature of a hypothetical microorganism that supposedly caused both scarlatina, smallpox, foot-and-mouth disease and syphilis. These fantastic theories go a long way to demonstrate the complete confusion at the time about the cause of syphilis.

Schaudinn was able to demonstrate a pale, spiral-shaped rod in a preparation from a fresh syphilitic skin lesion. He called the new microorganism *Spirochaeta pallida* (now renamed *Treponema pallidum* or Schaudinn's bacterium). On the other hand, similar spirochetes could be found in lesions that were not of a syphilitic nature, and when Schaudinn in 1905 reported his findings, he took care not to claim that *Spirochaeta pallida* caused syphilis. However, a few months later he could sustain this tentative conclusion when he demonstrated the presence of the spirochete in syphilitic lymph nodes. Nevertheless, there was considerable distrust among leading physicians in Berlin when he presented his discoveries to them, but independent research in other countries soon confirmed his results. Unfortunately, Schaudinn died soon after of sepsis.

A prerequisite for a systematic investigation of a great number of compounds as possible remedies for a disease is to have a suitable animal model that can be used for the testing. It was therefore a great step forward in the attempts to find a cure for syphilis when Émile Roux and Elie Metchnikoff at the Pasteur Institute in 1903 were able to infect monkeys with syphilis. Ehrlich also started to work with monkeys in his search for atoxyl derivatives that were effective not only against trypanosome illnesses like sleeping sickness, but also hopefully could be used in the treatment of human syphilis. However, the cost of using monkeys in such tests was prohibitive, and when Sahachiro Hata arrived from Japan in 1909 in order to work with Ehrlich on spirochete-induced diseases, he began to use relapsing fever in rats and mice as well as syphilis in rabbits as animal models.

Hata now tested with all the patience of the Orient, the numerous arsenic compounds that had accumulated in the laboratory during many years of trypanosome research and that had been well tolerated by the laboratory animals. Eventually he arrived at a preparation, a distant relative of atoxyl, dihydroxy-diamino-arsenobenzene, that gave a soluble salt with hydrochloric acid and had been given the preparation number 606. When Hata tested this compound it was well tolerated and effectively cured not only relapsing fever in rats and mice, but also syphilis in rabbits.

In the beginning, Ehrlich was somewhat guarded in his attitude to 606. He had seen too many compounds that had seemed promising enough but in the end had to be rejected because the therapeutic effect was not certain enough or because of unpleasant side effects. However, preparation 606 seemed really promising, and he became more and more optimistic when he saw the results of the therapeutic experiments on animals and the extensive toxicity tests. The demand for 606 was such that the Höchst Company had to work long hours to provide Ehrlich with the product of the final but one step in the synthesis of 606. The very last step in the synthesis was done in Ehrlich's own laboratory and was a rather tricky one since it had to be carried out under anaerobic conditions (i.e. free of oxygen). The final product, the 606 itself, was very sensitive to oxidation, which gave poisonous products, and it therefore had to be kept in vacuum-sealed glass ampoules.

For the clinical testing of 606, a total of 65,000 doses were manufactured in the Georg-Speyer-Haus and provided *gratis* to a selected number of physicians, whom Ehrlich had confidence in. He constantly worried about how the sensitive compound would be stored and whether it might be administered to the patients in an incorrect way if it were made freely available. At least on one occasion, his worries proved to be well founded. The reports of the doctors, who participated in the clinical testing, were on the whole positive, not to say enthusiastic. There was one exception, however, the dermatological clinic in Prague where severe complications, involving the kidneys and the nervous system, had been observed after injections of 606. Ehrlich was not satisfied until he had been able to show that the ampoules used on these occasions had been opened and then resealed several days before the content had been injected into the patients. This was contrary to his strict instructions for how 606 should be handled and here, in Ehrlich's opinion, we had the obvious explanation of the complications. He may well have been right and in any case the incident illustrates how very thoroughly he followed the clinical testing of 606. To summarize, the results during the first year 1909/10 were very good both for relapsing fever and fresh syphilis, while the effect on syphilis of long standing, for instance paralytic cases, was not encouraging. On the whole, it seemed well justified when Ehrlich and the directors of Höchst decided to rename preparation 606 as *salvarsan* (approximately "arsenic that cures").

Salvarsan — Success and Trouble

Ehrlich may well have believed that he would eventually be able to delegate both the production and the clinical testing of salvarsan to others, once the groundbreaking work had been done, so that he could devote his time to new projects. However, this proved to be a naive hope. The very success of the treatment and the enormous demand for the new drug all over the world, made him the prisonor of salvarsan for the rest of his life. Of course, Ehrlich was far from being insensitive to the fame that he now enjoyed as the greatest living figure in microbiology after the demise of Robert Koch. Even his old friend (and sometimes adversary) Emil von Behring, the only one that might challenge his eminence in the field of biomedical research, joined in the homage and seemed to have forgotten that he had once tried to dissuade Ehrlich from concentrating on the development of chemotherapy. In a long and warm letter to Behring, written in the summer of 1910, Ehrlich thanked him for his congratulations and deplored the long period of illness that had hindered Behring's own research. It is obvious that the revival of his old friendship with Behring, after a period of somewhat cooler relations, had been a great joy to him. However, unavoidably, there were also certain jarring sounds amidst all the praise.

While the ovations of the press knew no boundaries (one paper asserted that Ehrlich with Jesus Christ were the greatest figures of the Jewish people), some of his medical colleagues were instead considerably more restrained. They pointed to the side effects of the salvarsan treatment and even maintained that the old mercury therapy might after all be the best method of treatment. Others complained of not having been allowed to participate in the clinical testing of salvarsan. In many cases, the criticism stemmed from professional envy, sometimes perhaps also colored by the anti-Semitism that lay latent during this period, but would a few decades later culminate in a dreadful massacre under the Nazi regime. A dermatologist and syphilis specialist as well as police physician, Dr. Dreuw, was particularly unrelenting in his attacks on Ehrlich, whom he accused of neglecting the toxicity testing of salvarsan and running an advertising campaign in the press. These unfair and false accusations made Ehrlich and his followers very upset. One of Ehrlich's relatives even succeeded in having the malefactor removed from his position as a police physician, which hardly succeeded in endearing Dr. Dreuw towards Ehrlich and his salvarsan.

In the spring of 1914, a grotesque episode occurred that led to the so-called salvarsan action. In Frankfurt am Main there lived an, until then, rather harmless eccentric by the name of Karl Wassmann, who in the garment of a monk, was in the habit of trying to peddle in the streets of the town, a periodical produced by himself with varying names, for instance, "The Truth", "Love" or "The Free Spirit." In his journal, he advocated, among other things, natural healing, but now he suddenly asserted that in the Frankfurt Hospital, under the direction of professor Herxheimer, prostitute patients were forced against their will to undergo painful and perilous treatments with salvarsan, which had caused incurable injuries. Wassmann claimed that this had been done for research purposes, and also to increase the sale of salvarsan and profits of the Höchst Company and Ehrlich himself.

No one paid any attention to Wassmann's effusions, except professor Herxheimer who became terribly upset. With the aid of a young and ambitious prosecutor, he brought on legal action against the dotty Wassmann for defamation, despite the fact that Ehrlich made every effort to prevent the proceedings. To his despair, Ehrlich had to appear as a witness in court; it was hardly any consolation to him that the defendent was eventually convicted and sentenced to a year in prison. In any case, the salvarsan action infused fresh life into all the old slander, not least the assertion of immense incomes from the sale of salvarsan, and a preposterous rise in value of the Höchst shares. In the end, Ehrlich felt that he had to respond to the public allegations, thus he wrote a long article in the *Münchner Medizinischen Wochenschrift*, where he explained in a very didactic manner the real costs of producing and testing a new drug such as salvarsan. In spite of Ehrlich's difficulties of expressing himself in writing even in his native German (he always envied Behring his facility as a writer), this article was a truly admirable achievement.

The fanaticism of his salvarsan adversaries and their spiteful and irrelevant arguments must not disguise the fact that there were real problems with the salvarsan, which cannot be explained by incorrect storing or administration. To begin with, the usefulness of the drug was limited, it was effective only on fresh syphilis. Furthermore, subcutaneous or intramuscular injections of salvarsan caused severe pain in some patients, while others experienced no serious discomforts. However, the greatest problem was that salvarsan sometimes caused inexplainable deaths, perhaps due to some hypersensitivity of the patients. Until 1914, hundreds of thousands of treatments with salvarsan were given all over the world, and during this time approximately 300 serious injuries or deaths that could be attributed to the drug were reported. The mortality would thus be somewhere around 1/1000, a reasonably low figure considering that it was a question of treating a very serious illness for which there was no other cure. Even so, Ehrlich was worried about the mortality and he tried to produce a less toxic derivative of salvarsan. Eventually, Alfred Bertheim, one of the chemists at the Georg-Speyer-Haus, managed to synthesize a derivative that was easily soluble and seemed to give less complications on injection. It was released for clinical use in 1912 under the name of neosalvarsan and was generally thought to represent a significant improvement.

Honors and a Premature Death

Paul Ehrlich had always shunned publicity and preferred a quiet fame among his scientific colleagues, where he undoubtedly had become a great name, not least internationally. Even so, he had received a number of honors, among them the Nobel Prize in Physiology or Medicine of 1908, a couple of years before the great breakthrough with salvarsan. At the time, other honors were probably equally important. In 1911, the official Prussia showed its esteem by appointing him *Wirklicher Geheimrat* with the title *Excellenz*, the same position that had earlier been granted to Koch and Behring and the highest distinction that the Prussian government could bestow on any scientist. He had also received countless medals, for instance the great Prussian gold medal for science and the very prestigious Liebig medal, not to mention a number of honorary doctor's degrees and an assortment of different decorations. To celebrate his 60th birthday, a so-called *Festschrift* was published, where his great achievements were celebrated in 37 chapters. Certainly, this must have been rather gratifying even to the publicity-hating Ehrlich!

His last years were darkened by the outbreak of the First World War, which drastically curtailed his international scientific contacts. When he heard the shattering news about the war, Ehrlich is said to have cried out in despair: "This is pure lunacy!" He was undeniably much more clear-sighted here than the crowds that greeted the declaration of war with patriotic cheers. The enormous workload that the salvarsan had entailed was of course also a strain on his health, which was hardly improved by his constant cigar smoking, a habit that he had acquired in his youth. In 1914, his health deteriorated significantly and he showed increasing signs of circulatory discomfort. At Christmas time, he suffered a minor stroke, but his condition seemed to improve, even if he missed his beloved, albeit forbidden cigars. In August 1915, Ehrlich entered a nursing home to recuperate, and here he suffered another stroke. He died peacefully on 20 August without having regained conciousness.

At his funeral in the Jewish cemetery in Frankfurt am Main, his friend Emil von Behring said in a moving eulogy, that the deceased had become a *Magister Mundi*, a teacher of medical science all over the world. Perhaps, this was Paul Ehrlich's greatest role, as an intellectual leader and inspirer who directed the medical thinking on new and fruitful paths. Today, scientists worldwide certainly look for molecular explanations for all sorts of biomedical problems, and without a doubt, Paul Ehrlich was the man who really opened our eyes to this new line of thought.

ELIE METCHNIKOFF

Since the middle of the 19th century, medical thinking had been dominated by Virchow and his cellular pathology. At the end of the century, when microbiology had its great breakthrough in medicine, the relations between Rudolf Virchow and the Koch school of bacteriology had been far from friendly. On the other hand, Julius Cohnheim, another pioneer of modern pathology, had an entirely different attitude, and actively encouraged both Koch and Ehrlich. On the whole, however, there was a contrast between the cellular thinking of Virchow's pathology and the new idea of antibodies, freely circulating in the body fluids, as the basis of immunity. This was also emphasized by their discoverer, Emil von Behring, when he talked about going back to the old Hippocratic principles of humoral pathology. As the main protagonist of the opposite standpoint, what came to be called cellular immunity, appeared Elie Metchnikoff. Nevertheless, he and Behring still became good friends in the end.

A Sensitive Young Man

Elie Metchnikoff was born in 1845 as the youngest of five children. His father, Ilia Ivanovitch, was a guards officer who wasted away most of his and his wife's fortune. He was later forced to leave St. Petersburg for retirement to his estate in the neighborhood of Charkov. Ilia Ivanovitch did not seem to have any intellectual interests; his only great passions in life were card-playing and gluttony. Elie's mother Emilia, daughter of the Jewish writer Leo Nevahovitch, was on the other hand a high-spirited and intelligent woman with considerable cultural interests, who played a major role in the boy's academic development and choice of career. As a child, Elie was sensitive and spirited, almost restless — in the family he was known by the nickname "Mercury." With his charm and spontaneity, he was his mother's favorite. His brothers and sisters however, considered him to be a spoiled and tiresome brat. A private tutor, who had been engaged to instruct one of Elie's elder brothers, became interested in the precocious eight-year-old Elie — taught him natural sciences and took him on excursions to collect and examine flowers. He soon became an expert on the local flora; and when at the age of 11, he was admitted to the Gymnasium in Charkov, his lifelong interest in biology had already been established.

Several of the teachers at the Gymnasium were comparatively progressive pedagogues who used modern educational principles. Elie seems to have enjoyed his schooldays and was an outstanding student, even if he tended to skip subjects that he found boring and instead concentrate on the natural sciences. At the age of 17, he passed the final examination at the Gymnasium with honors and was rewarded with a gold medal — a distinction which convinced his parents that the

Elie Metchnikoff (1845–1916).
Courtesy of the Nobel Foundation, Stockholm, Sweden.

boy should be allowed to continue his studies at the university. During his schooldays, Elie had followed some of the courses at Charkov University, but had not been favorably impressed by the education there, which he found old-fashioned. At the German universities, on the other hand, the quality of both research and education was much higher, something that Elie was well aware of. He was determined to study biology at Würzburg where the eminent zoologist Rudolf von Kölliker was active. Emilia helped her son persuade his father to finance the project and so the young student embarked on his ardently desired trip abroad. Unfortunately, it proved to be full of misadventures and disappointments. Worst of all, it was not until he arrived at Würzburg that he discovered that the term only started six weeks later. Alone and unhappy in a foreign city, he was seized with an acute attack of homesickness and immediately caught the next train back to Russia and his family. His first excursion into the world of international science had ended somewhat ignominiously, but later attempts would turn out considerably better.

Elie then decided to remain at Charkov University to pursue medical studies. However, Elie's mother did not think it was a good idea and tried to dissuade him. In her opinion, he was much too sensitive to put up with all the human suffering that he would be exposed to as a medical student. Considering some of the things that happened in coming years, she may well have been right here. Metchnikoff thus spent two years in Charkov studying biology, but he still yearned for the world outside his hometown, far from the confined atmosphere of a Russian provincial university. In 1864, he finally had the chance to spend two months on the island of Helgoland outside the German North Sea coast, where he studied marine biology. His parents willingly supported him financially, and this time his trip abroad was a success. He made the acquaintance of the famous botanist Ferdinand Cohn, who was impressed by Metchnikoff's ardent love of research and advised him to contact the zoologist Rudolf Leuckart in Giessen. Here, Metchnikoff made some interesting observations regarding the propagation of the nematodes, but his intense work at the microscope affected his eyes. He began to suffer from a severe eyestrain, and eventually all microscope work became impossible. His eye condition continued to trouble him periodically for many years.

Metchnikoff had got into a conflict with Leuckart over the publication of some results they had obtained together, but thanks to a stipend for two years from the Russian Ministry of Education, he could leave Giessen and go to Naples where the opportunities for marine biological research were incomparably better. Strongly influenced by Darwin's ideas about evolution, Metchnikoff was of the opinion that embryos from lower organisms (invertebrates) developed according to the same principles as vertebrate embryos. In Naples, he collaborated with a kindred spirit, the young Russian zoologist Alexander Kovalevsky, who also became a close friend. Together, they contributed to the appearance of a new branch of science, comparative embryology. It was not until 1867 that Metchnikoff, together with Kovalevsky, returned to Russia where he got his doctor's degree at the university of St. Petersburg, and in addition was honored with the prestigious von Baer Prize for his research on the origin of the germ layer of invertebrate embryos. He was also offered a position at the university of Odessa. At the age of 22, Elie Metchnikoff was already considered an established personality in the scientific world.

However, his academic career in Russia was destined to be full of disappoint- ments and conflicts. Perhaps, this was not entirely the fault of reactionary and obdurant colleagues. The young Metchnikoff was not always very diplomatic and he may have had a tendency to dogmatism. It did not take long before he quarrelled with an old and scientifically not very outstanding professor about who should have the distinction of representing the university at a congress of natural scientists in St. Petersburg. In the end, they were both chosen as delegates to the congress, but the conflict had been bitter and did not endear Metchnikoff to the senior members of the faculty. He also found his laboratory and working conditions generally very unsatisfactory. Life as a professor in Odessa seemed increasingly unbearable to him, in spite of the fact that he was highly regarded by his students. Thus, when Metchnikoff was offered a chair in zoology at St. Petersburg University he was not difficult to persuade. A new trip to Naples with Kovalevsky was not as scientifically successful as the first one; he had trouble with his eyes and worried that this might mean the end of his scientific career. Furthermore, the opportunities for research in St. Petersburg turned out to be even worse than in Odessa.

The following years must have been the most difficult in Metchnikoff's life. In 1869, he married the young Ludmilla Federovna — a step that deeply worried his mother. Ludmilla suffered from pulmonary tuberculosis, and despite the heavy economic sacrifices that Metchnikoff made in order for her to stay in a warmer climate, her condition continued to worsen. His position in St. Petersburg made it necessary for man and wife to live apart for long periods. The last year of her life, Ludmilla stayed on Madeira. When Metchnikoff saw his wife again in the spring of 1873, after having been absent for several months, she was already dying. Though her death did not come as a surprise, Metchnikoff was deeply depressed. On the journey home, Metchnikoff visited his brother Leo in Geneva. During his stay there, he attempted suicide by swallowing a large dose of morphia, which thankfully did not kill him, instead only induced violent vomiting.

Metchnikoff had recently been offered a new position at his old university in Odessa, but after his wife's death and his return to Russia he did not think that he would be able to resume his biological research. His eyes troubled him more than ever and he could not work at the microscope. Instead, he decided to devote himself to anthropology. He had in mind to study the Mongol Kalmucks, who inhabited the Astrakhan steppes, and their development from early childhood to adult age. Obviously, he saw this as a continuation of his previous work in comparative embryology.

There were a number of problems in connection with the expedition. The money for the expedition had to come out of his own pocket, he did not know a word of the Kalmuck language, travelling conditions were primitive to say the least, everything was indescribably dirty, and the food consisted mainly of mutton that stank of rancid fat. Nevertheless, Metchnikoff was not discouraged but worked indefatigably with his anthropological measurements, making several interesting observations, at least in his own opinion. It seemed to him that the physical development of the adult Kalmucks had remained on an infantile stage compared to the Caucasians: a big head, short legs and a dominating torso — exactly as in Russian infants. He also had an explanation of this apparent inhibition of development in the Kalmucks. It was due to a chronic intoxication caused by their constant consumption of fermented milk. A daring hypothesis, to be sure, and hardly one of Metchnikoff's more important contributions to science.

Having returned to Odessa, Metchnikoff lived in a flat just beneath that of a large family of eight rather lively children, whom he found a bit disturbing, particularly in the morning when he wanted to sleep late. Later, he became interested in the eldest girl, 16-year-old Olga Belokopitova, whom he offered to give lessons in natural science. It did not take long before teacher and pupil were head-over-heels in love and in February 1875 they were married. The wedding took place on a snowy winter day; Olga in her biography of Elie Metchnikoff relates how her younger brothers persuaded her to sledge with them just before the solemn ceremony. Everyone had a marvellous time, with the exception of the mother of the bride who had the shock of her life when she found out what her daughter was up to. The episode certainly illustrates the youth, not to say childishness of the bride, but even so their marriage would last a lifetime and Olga became a loving support of the somewhat unstable Metchnikoff and an indispensable collaborator in his research.

The first year of their marriage coincided with a period of political unrest in Russia, characterized by opposition among the intellectuals and the students against the government's oppression, and by terrorist acts committed by desperate anarchists. Metchnikoff was far from being a revolutionary but he, nevertheless, found the reactionary tsarist regime unbearable — and it would get worse. In March 1881, the anarchists murdered the relatively liberal tsar Alexander II. Thereafter, during his brutal successor Alexander III's reign, the oppression intensified. The working conditions at the university were such that both research and teaching became impossible for Metchnikoff, who was increasingly forced to take sides in the conflict between conservative and liberal professors. To make things worse, his young wife had fallen very ill with paratyphoid and he had feared for her life. Though she recovered, Metchnikoff still entered into a state of depression, and in 1881 attempted suicide a second time by injecting himself with a culture of the spirochete that causes relapsing fever. He became seriously ill but eventually recovered, though still suffering from persistent heart problems which he blamed on the bout of relapsing fever. By this time both Olga's parents had died, thus Metchnikoff had to assume the responsibility of guardian to her younger brothers and sisters. The inheritance left behind by Olga's father allowed the family to be economically independent, and it became possible for them to leave Russia and settle abroad where the conditions for his research were much better.

The Road to Paris

In the autumn of 1882, the whole family including five of Olga's younger brothers and sisters set out for Messina, and with that Metchnikoff had taken the first step on the road that would ultimately lead to Paris and the Pasteur Institute. He loved Messina and the Mediterranean with its rich marine fauna, where he had made his first important discoveries. He had early on observed that cells from marine organisms could take up dye granules in water suspension and similar findings had been reported by the German biologist Ernst Haeckel. Metchnikoff now began to study this process systematically in the transparent larvae of starfish. He found that mobile cells from the mesoderm of the larvae gathered around foreign bodies. Metchnikoff was struck by the thought that this could be a phenomenon similar to the inflammatory reaction that followed when you got a thorn in your finger. He made experiments where he run thorns into his larvae and sure enough could see that cells gathered around the thorns. Obviously, this was yet another proof of his favorite hypothesis of a correspondence between primitive organisms and higher species with a developed vascular system.

In 1884, Metchnikoff made another important observation when he studied the water flea Daphnia. The intestinal canal of these animals often contained the elongated, pointed spores of the yeast *Monospora bicuspidata*. Occasionally, the peristaltic movements made the pointed spores penetrate the intestinal wall into the abdominal cavity of the flea. In some cases, the resulting infection killed the animal, but usually the flea survived. It turned out that the flea contained a defence mechanism in the form of mobile cells which engulfed the spores and degraded them. Some years later, Metchnikoff found that another parasite (*Pasteuria ramosa*), which always killed the infected Daphnia, was not taken up and degraded by the mobile cells of the flea.

Metchnikoff became increasingly convinced that cells with the ability to take up and digest foreign bodies (he called this phenomenon phagocytosis) had an important role in the defence against invading microorganisms, not only in starfish larvae and water fleas, but also in the human organism. In his own words, this idea changed him from a zoologist into a pathologist. Metchnikoff would, in coming years, devote most of his scientific career to developing and defending his theory of phagocytosis and cellular immunity. Rudolf Virchow, who happened to be in Messina in 1882 when Metchnikoff made his first discoveries of phagocytosis in the starfish larvae, encouraged him but at the same time pointed out that his theories about the role of the inflammatory reaction in the defence against bacterial infections, run contrary to the generally accepted view in pathology. In fact, the most famous of Virchow's pupils, Julius Cohnheim, had made pioneering studies of the inflammatory process and had shown that the leucocytes, which participated here, were derived from the circulating blood. He was therefore of the opinion that only animals with a vascular system were capable of this reaction. Furthermore, it was generally agreed that the leucocytes, by taking up bacteria, helped spread these in the infected organism. Phagocytosis, far from being a defence against microorganisms, instead made the infection worse. Thus, there was a lot of resistance to overcome before Metchnikoff's ideas could be accepted in medicine.

His first publications about phagocytosis and its importance for the defence against infections were released in 1883 and continued uninterrupted for a quarter of a century. Metchnikoff's fame as a biologist increased steadily, even if there was a certain scepticism against the medical importance of his discoveries. Inspired by Pasteur's work (that was now recognized all over the world), the authorities in Odessa had in 1886 established an institute for microbiological basic research that was to be combined with the production of rabies vaccine. Metchnikoff was offered the directorship of the institute and he took up his new position with the greatest expectations. However, internal conflicts with members of the scientific staff made his situation difficult. In addition, the authorities obstructed his immunological research on the pretext that he was not licensed to practise medicine. After only two years, he and Olga left Russia for good in search of a place in Europe where they could settle down.

Metchnikoff had visited Munich (where he conflicted with the conservative hygienist Rudolf Emmerich, who criticised his phagocytosis theory) and Berlin (where Koch to his surprise and disappointment had received him rather coldly). This was in contrast to Pasteur, who had commended Metchnikoff's research and generously offered him a laboratory in the new institute which was being constructed. The life in an international metropolis like Paris intimidated him to some extent. He would have preferred a quiet little university town, but at the same time he was very attracted by the atmosphere of the Pasteur Institute, whose famous leader was surrounded by an enthusiastic team of young scientists. In the autumn of 1888, Metchnikoff moved into a laboratory consisting of two rooms on the second floor of the new institute building. He and Olga had made their choice and they obviously never regretted this decision. France became their new home and Metchnikoff remained at the Pasteur Institute for the rest of his life. They seem to have liked the place right from the start, and of course their life was made more comfortable with the help of Olga's fortune in Russia which was enough to support them. During his 28-year stay at the Pasteur Institute, Metchnikoff did not receive any salary.

The Metchnikoffs (Olga functioned all this time as Elie's research assistant) were now members of the rather special group around Pasteur, a family of scientists that has sometimes been described as a monastic order completely dedicated to research. Elie's personal relations to the great man seem to have been the very best, not really a matter of course with a person like Pasteur. His younger colleagues soon also learned to value Metchnikoff for his vast knowledge of the literature and the generosity with which he put it at their disposal. Particularly Émile Roux, director of the Pasteur Institute since 1904, became a very close friend.

Metchnikoff soon succeeded in gathering around him a group of young scientists; among them the best known name is Jules Bordet, who in 1919 received the medical Nobel Prize for his discoveries in immunology. The medical community had been reluctant to accept phagocytosis and its alleged role in the organisms defense against infections, particularly after Behring's discovery of the antibodies and the therapeutic breakthrough that followed. However, Metchnikoff continued indefatigably to assert the importance of phagocytosis, and a good example is his attempt to explain the so-called "Pfeiffer's phenomenon" from the point of view of cellular immunity. Richard Pfeiffer, an outstanding pupil of Koch, had shown in 1894 that cholera vibrions injected intraperitoneally (i.e. through the peritoneum into the abdominal cavity) in guinea pigs that had previously been vaccinated against cholera, were killed within a few minutes and could be observed as motionless granules in the peritoneal fluid. On the other hand, serum from animals immunized against cholera had no effect on vibrions *in vitro* (i.e. in test tube experiments). However, Pfeiffer also made *in vitro* experiments where he mixed immune serum from vaccinated animals with peritoneal fluid from normal guinea pigs, and he found that this mixture killed and dissolved cholera vibrions. It seemed that the peritoneal fluid from normal animals contained a factor that was necessary in order for the immune serum to kill vibrions *in vitro*.

Pfeiffer had gone out of his way to point out that, in his opinion, phagocytosis had nothing to do with this phenomenon. Metchnikoff saw this as an attack on his cellular immunity theory and he immediately accepted the challenge. He had previously worked with cholera and had doubted then that Koch's vibrion had anything to do with the disease. He thus shared Pettenkofer's views, and like him conducted experiments where he swallowed live cholera vibrions without becoming ill. Encouraged by these results, he extended the experiment to include two other volunteering collaborators. Unfortunately, one of them became ill with typical symptoms of cholera. Metchnikoff was of course very upset and he must have been enormously relieved when the patient eventually recovered. From then on, Metchnikoff was completely convinced that the vibrions were the cause of cholera. The whole episode is typical of the rather careless attitude towards experiments on human beings during the infancy of bacteriology.

Metchnikoff repeated Pfeiffer's experiments, and although he got the same results, his explanation was entirely different. In his opinion, the destruction of the vibrions was not caused by antibodies but was due to leucocytes, damaged by the injection. These damaged leucocytes leaked intracellular enzymes (he called them "cytases") that digested the bacteria. Metchnikoff made a series of similar experiments, which seemed to confirm that the vibrions were destroyed by a diffusion of "cytases." As it turned out, his "cytases" were not enzymes but "complement," proteins that appear normally in the body fluids and together with the antibodies participate in the breakdown of foreign bodies like bacteria and non-compatible blood cells. Jules Bordet showed that normal serum contains these factors, which he called "alexin," not "complement," while he referred to antibodies as "substance sencibilatrice" (Metchnikoff preferred the term "fixateur" instead of antibody). There was an almost total terminological confusion in immunology at this time.

Metchnikoff had always been highly regarded as a lecturer, and in 1891 he conducted a survey of the different theories about the nature of the inflammatory process, which drew large audiences. The next year, it was published under the title: *Lecons sur la Pathologie Comparée de l'Inflammation* ("*Comparative Pathology of the Inflammatory Process*"). At the turn of the century, Metchnikoff started to write a comprehensive treatise in which he tried to summarize the whole field of immunology: *L'immunité dans les Maladies Infectieuses* ("*Immunity in Infectious Diseases*"). Here, he again defended his theory of phagocytosis and its importance for immunity, and he criticised Behring and Ehrlich for exaggerating the role of the free antibodies. The book was published in 1901 and created considerable interest. Nevertheless, it must be admitted that medical science tended to lose interest in cellular immunity during the first decade of the new century. On the other hand, looking back at the development of immunology since Metchnikoff's discoveries in the 1880s, their great importance has become increasingly obvious over the years. One must really commend the perceptivity of the medical Nobel Committee when they let Metchnikoff share the prize with Ehrlich in 1908.

Aging and Death

During the last 15 years of his life, Metchnikoff became increasingly fascinated by aging and death as biological processes. Why did degenerative phenomena such as arteriosclerosis increase with age and as a consequence also illnesses that are related to such changes? His interest in vascular diseases made Metchnikoff together with Roux look for an animal model of syphilis, an illness that also affected the vascular system, and which until then, had been considered an exclusively human disease. In 1903, they were able to infect monkeys with syphilis, an experimental breakthrough that was important also for Ehrlich's work with salvarsan, even if for economic reasons the German group eventually had to choose another animal model.

In order to postpone the inevitable aging as much as possible, Metchnikoff relied mainly on a generally wholesome way of life. In particular, he stressed the importance of a diet that prevented the growth of harmful bacteria in the intestines with the formation of bacterial waste products that led to a chronic poisoning of the whole organism. At the end of a lecture in Manchester in 1901, he declared that the intestinal bacteria are the main reason for the shortness of the human life. He also expressed the hope that during the new century science would be able to solve this problem, thus increasing the human life span. Personally, he put great stock in a diet of yoghurt that would lead to a colonization of the colon by lactobacilli which would prevent the growth of harmful bacteria. Here, he could point to people in the Caucasus and the Balkans, who are said to live to a ripe old age because of this diet.

Another reason for the early senility that shortened human lives was alcoholism and infectious diseases, in particular syphilis. If people could only be persuaded to live according to the healthy principles laid down by Metchnikoff, he called it orthobiosis (right living), their life span would be dramatically increased.

Furthermore, science might be expected to eradicate all infectious diseases, the dominating cause of death at this time. The fear of death would then disappear, when it was just perceived as the peaceful termination of a life that had reached its completion.

Of course, Metchnikoff himself led an orthobiotic life, and in 1913 he could triumphantly write in his postumous notes that he had successfully lived longer than either his parents or brothers and sisters. It is true that he had trouble with his heart (at the time he suffered a minor coronary), but in Metchnkoff's opinion this was because it was only at the age of 53 that he started to lead an orthobiotic life. With typical 19th century optimism, he believed that the steady progress of science would lead to a better future for man and give humanity not only health and prosperity but also make us more sensible and increase our moral standing. This naive belief in a future of reason and peace was shattered by the outbreak of the First World War. They were then living outside Paris and Metchnikoff had to take the train to the Pasteur Institute everyday. Olga relates in her biography how Elie arrived at the institute after the declaration of war and found that it had been completely taken over by the military authorities. The laboratories had been emptied of all the young staff and basic medical research was not possible any more. She says that it was an aged and broken man that got off the train from Paris that evening when she as per usual came to the railway station to meet him. The war years, with their terrible loss of human lives, among them many of his friends and collaborators, and his enforced scientific inactivity certainly contributed to the steady decline of Metchnikoff's health. At the end, he was moved on the initiative of Émile Roux to what had once been Pasteur's private apartment at the institute. The grievously ill Metchnikoff saw this as a great honor, which made him very happy. And it was here that he died of heart failure on 15 July 1916.

NOBEL PRIZES AND NOBEL COMMITTEES

1901

The awarding of the first Nobel Prizes in 1901 had been preceded by a series of conflicts and difficulties, which had started with the death of Alfred Nobel in 1896. His young collaborator, Ragnar Sohlman, had to smuggle the assets of the Nobel estate out of France in the greatest secrecy to prevent them from falling into the hands of the French authorities. This had to do with the question of Nobel's legal residence at the time of his death, which may have affected both the size of the death duties and possibly also the validity of his will. Even after the estate had arrived safely in Sweden, there remained a number of serious problems that had to be solved before the prizes (which Nobel had provided for in his will) could be realized. Some of his relatives, for instance, did their utmost to have the provisions of the will nullified. In addition, there were controversies on what Nobel had really intended with some of the stipulations in his will.

For example, the prize for literature was supposed to go to an author, who during the preceding year had produced the best work "in an ideal direction," whatever that might mean. Even the scientific prizes in physics, chemistry and physiology or medicine, presented problems since the will stipulated that the prizes should go to scientists "who had during the preceding year done humanity the greatest service" in their respective fields of research. The benefit of a certain scientific discovery for humanity has always been difficult to assess and even if, to make things as simple as possible, we take "benefit for humanity" to mean "scientific quality," the incomparably greatest difficulty still remained — the time.

It often takes considerable time before it is possible to judge the importance of a scientific discovery. In some cases, it can take decades before the true dimensions of a discovery is fully realized. In any case, to really estimate the significance of a discovery that was made the previous year is out of the question. This is of course equally true of the prize for literature, not to mention the peace prize, where a number of other, apparently unsurmountable obstacles are added. Strangely enough, it turned out that the Norwegian "Storting" (parliament), which Nobel had selected to award the peace prize, was the only prize-awarding institution that created no difficulties, and immediately accepted the task. The other institutions supposedly custodians of the Nobel Prizes, for instance the Royal Swedish Academy of Sciences for the prizes in physics and chemistry, and the Karolinska Institutet for the prize in physiology or medicine, had that many more objections. With the original provisions of the will, they refused to take responsibility for selecting the scientific Nobel laureates and the Swedish Academy had the same attitude regarding the prize for literature. The great project seemed to have arrived at an impasse.

Alfred Nobel (1833–1896).
Courtesy of the Nobel Foundation, Stockholm, Sweden.

It would take all the ingenuity, devotion and indomitable willpower of Ragnar Sohlman, whom Nobel had appointed as one of the executors of his will, to overcome these difficulties. To begin with, some of the more rigid provisions in the will of Nobel had to be made more flexible before the Swedish government could confirm the charter of the Nobel Foundation. The evaluation of the candidates had now been made easier by stipulating that the prize could be divided between a maximum of three laureates. Furthermore, in an important paragraph of the charter, it was stated that work done before a year ago can also be rewarded "in case its importance has not been realized until lately." With this useful and rather flexible clause in place, both the Academy of Sciences and the Karolinska Institutet found that they could assume the responsibility for the scientific prizes, while the Swedish Academy became in charge of the prize for literature.

Nobel Committee and Nominees

The faculty of the Karolinska Institutet had in 1901 selected the following Medical Nobel Committee to evaluate the candidates for the first Nobel Prize in physiology or medicine:

Almquist, Ernst	Professor of Hygiene
Henschen, Salomon E.	Professor of Medicine and Pathology
Müller, Erik	Professor of Anatomy
Mörner, Karl A. H.	Professor of Medical Biochemistry (chairman)
Sundberg, Carl	Professor of Pathological Anatomy (all from the Karolinska Institutet)
Tigerstedt, Robert	Professor of Physiology at Helsingfors University, Finland

The Karolinska Institutet had invited a number of internationally recognised scientists and scientific institutions to suggest candidates for the prize, a procedure that is still in use. At the end of the stipulated time period, the committee found that Emil von Behring had received 13 nominations (in one case together with Shibasaburo Kitasato, and in another together with Paul Ehrlich) for the discovery of the antibodies. It is worth noting that no less than nine of these nominations came from the medical faculty of Leyden University. Robert Koch had received four nominations for his work on malaria and Elie Metchnikoff had three for his discovery of phagocytosis.

Ragnar Sohlman (1870–1948).
Courtesy of the Nobel Foundation, Stockholm, Sweden.

The Discovery of the Antibodies

The committee had asked its member, Ernst Almquist, professor of hygiene, to make an evaluation of Behring's great discovery and it must be said that the memo, which Almquist submitted to his colleagues, is remarkable in several respects. To begin with, one is struck by how short it is, only three typewritten pages, and by the generally poor impression it makes. The author dwells in relative detail on trivial things like the difficulties that Behring had experienced in finding money for maintaining the number of laboratory animals required for his immunization experiments. He also discusses the possible importance of Koch's work on tuberculin as an example and inspiration for Behring, but concludes quite correctly that "this discovery is undoubtedly in essence Behring's own." He also finds that Kitasato's contribution to the original project is considerably smaller than that of Behring. Almquist gives a short review of possible predecessors of Behring, but finds rightly that their names are not really important in this connection.

What is totally lacking in his memo is any real appreciation of the extremely important breakthrough in immunology that the discovery of the antibodies implies. He concludes his analysis by referring to a statement by Roux and Martin in 1894 to the effect that they on the whole are able to confirm the results of Behring and his collaborators. However, they also point out that the introduction of the serum therapy against diphteria might take more time than the discovery itself. These are hardly very enthusiastic concluding remarks and Almquist also leaves the question of Behring's merits as a laureate unanswered.

As an appendix to Almquist's memo there appeared another by Carl Sundberg, professor of pathological anatomy, that summarized the statistics about the therapeutic results of the serum therapy against diphteria until 1895. The author is convinced of the clinical usefulness of the serum and he agrees with a recent statement by Behring: "Already at the end of 1895 both the therapeutic effect and the harmlessness of the serum against diphteria were obvious to every physician who had any understanding of these matters." However in his memo, Sundberg does not address the question that must be answered: Is Behring the leading candidate for the Nobel Prize in physiology or medicine of 1901?

In this context, it is illuminating to take note of what Almquist and Sundberg have to say in a short evaluation of the Danish veterinarian Bernhard Bang and his efforts to combat cattle tuberculosis. In conclusion, they say the following: "We are of the opinion that this systematically planned organization, with its great results, might very well be rewarded with a Nobel Prize. But since a much greater discovery has been made in the last few years, which has also been nominated for the prize, it seems obvious to us that the preference must be given

Karl A. H. Mörner (1854-1917).
Courtesy of the Karolinska Institutet, Stockholm, Sweden.

Ernst Almquist (1852–1946).
Courtesy of the Karolinska Institutet, Stockholm, Sweden.

Carl Sundberg (1859–1931).
Courtesy of the Karolinska Institutet, Stockholm, Sweden.

to this latter discovery." The cryptic reference to a more worthy candidate cannot possibly allude to Behring, whose discovery had been made ten years earlier and could not by any stretch of imagination be said to fall within "the last few years." On the other hand, the description fits Ronald Ross and his discovery of the malaria vector marvellously; a candidate that the authors had also evaluated for the committee.

The Recommended Candidate

The memo entitled "The Malaria Question," which Almquist and Sundberg jointly submitted to the Nobel Committee deals mainly with the British military medical officer Ronald Ross and his discovery of the role of the Anopheles mosquito as a vector for malaria. Unlike Almquist's evaluation of Behring, this memo is rather comprehensive (more than six typewritten pages) and its judgements are extremely appreciative. Its closing remarks are as follows: "During the last years, Ross has made a pioneering discovery of the greatest practical and scientific value. In doing this he has demonstrated perseverance, originality and the greatest skill in the use of experimental methods. His discovery has been confirmed and acknowledged as the truth by the most outstanding scientists and it concerns one of the most important diseases. We therefore recommend that he be awarded a prize."

Comparing their evaluation of Ross with what they wrote about Behring and also what they said about Bang, one cannot doubt that Almquist and Sundberg must have recommended Ronald Ross instead of Emil von Behring for the prize. Today, with the benefit of hindsight, it is obvious that the discovery of the antibodies has been far more important for the development of modern medicine than the elucidation of the mosquito vector for malaria. Furthermore, there is every reason to believe that the majority of the medical world was of the same opinion even in 1901. The explanation of Almquist's and Sundberg's surprising suggestion must be that they had become enthralled by the idea that a discovery made during the last few years was always preferred to one that was made ten years ago. Undoubtedly, the provisions of Nobel's original will could have had an important psychological role here.

How could the final outcome be so completely different from what the two experts had suggested? Did they meet with opposition already in the committee, or did the other members go along with Almquist and Sundberg, but the faculty took a different view and settled for Behring? Both the deliberations of the committee and its recommendation to the faculty are usually oral and minutes are not taken, and the same is true of the discussion that precedes the final decision in the faculty. Nevertheless, we believe we know the answer to these questions. It appears that the committee had suggested that the faculty should award the prize with one half to Ronald Ross for his work on malaria and the other half to the Danish physician Niels Finsen for his phototherapy of skin tuberculosis. Ross got the prize the next year and Finsen in 1903, but this time the faculty rejected the committee's recommendation and decided instead on Emil von Behring. He was awarded the Nobel Prize in physiology or medicine of 1901 "for his work on serum therapy and especially its use against diphteria." Looking back at this decision a century ago, we can feel quite certain that it was the right one, and that Behring was in every respect a worthy laureate.

1905

Nobel Committee and Nominations

The medical Nobel Committee for 1905 had the following members:

Almquist, Ernst	Professor of Hygiene
Edgren, Johan G.	Professor of Medicine
Holmgren, Emil	Professor of Histology
Mörner, Karl A. H.	Professor of Medical Biochemistry (chairman)
Sundberg, Carl	Professor of Pathological Anatomy (all from the Karolinska Institutet)

Robert Koch already had four nominations in 1901 and the number had increased steadily every year until it had reached 20 in 1905, which made him the leading candidate in terms of number of nominations. One wonders why he had never won the prize all these years. After all, it was Koch, who together with Pasteur, had opened a new world to medical science. It was the work of Koch and his collaborators that had made the incredibly rapid progress of bacteriology possible. Why was he not awarded the Nobel Prize in 1901; who could possibly have been a more worthy candidate, if we consider his total contribution to medicine? The reason is, of course, the uncertainty of both the nominating scientists and the Nobel Committee regarding how old the discoveries could be, in order to be considered as merits. The rules said that discoveries made more than a year ago could be included, provided that their importance was only recently fully realized — but surely there must be some limit to how far back in time one could possibly go! When one reads the nominations and evaluations of candidates for the first Nobel Prizes, one gets the feeling that approximately ten years had become something of an arbitrary and never clearly defined borderline.

In the case of Robert Koch, it is obvious that the nominating scientists may as a matter of form have restricted themselves to his work on malaria during the last few years, but that in reality it was his total contribution to medicine that motivated their nomination of him. Elie Metchnikoff's nomination of Koch as early as 1901 is a good example of this, and professor Langhans from Bern was on the same occasion even more explicit, when he elaborated on why he had suggested Behring instead of Koch: "It goes without saying that Koch would have been the first choice, had it been possible to take his earlier work into consideration." The calamity with the tuberculin was of course embarrassing to the great pioneer, but it was the arbitrary time limit that created the real problem.

From 1903 and onwards, the proposers clearly felt more free to take also earlier work into consideration. In an exhaustively motivated proposal of Koch from 1903, professor Hoffmann from Leipzig emphasized Koch's pioneering work on anthrax and its transformation into a spore form, the classification of the bacteria of the infected wound and the identification of the tubercle bacillus — work that was now more than 20 years old. Generally speaking, Hoffman's nomination is a model of clarity and well balanced arguments.

Evaluating a Monument

In their memo "The Malaria Question" of 1901, Almquist and Sundberg mention only in passing Koch's work on malaria, in particular his attempts to limit the spreading of the disease. In a new memo on malaria of 1902, the year that Ross got the Nobel Prize, the authors again dwell mainly on the prophylactic work of Koch. It is not until in 1903 when Almquist and Sundberg have been commissioned to write a special memo about Robert Koch that they first mentioned about his pioneering bacteriological work in its whole extent, but even so their main emphasis is still on his work of the recent years. They concentrated on his malaria research and in particular his efforts to eradicate the disease in the German colonial empire. Furthermore, his more recent work on tuberculosis is described in considerable detail, with emphasis on cattle tuberculosis and the possibility of transmitting to humans. Koch had always been a sceptic here and instead had stressed the importance of, for instance, human expectorations. The authors also take up his methods for the diagnosis of typhoid fever and for limiting the epidemics that still occurred in Europe at this time.

Their concluding remarks are very revealing. One can see how the authors extremely cautiously begin to entertain the idea (rather obvious in our view) that Koch's latest studies can only be fully appreciated against the background of his previous groundbreaking work, that created the necessary conditions for his more recent research. Maybe they must after all regard Koch's discoveries and methodological achievements during the 1870s and 1880s as an important part of his merits for the Nobel Prize? Nevertheless, the emphasis is still on the research of the last few years, since in their opinion "Koch's older work cannot, in view of its age, be considered alone." A Nobel Prize must therefore be based above all on his recent epidemiological studies of malaria, typhoid fever and cholera, rather than on his having laid the groundwork for an entirely new science.

It is time for a new memo on the following year, but now Almquist is the sole author. It hardly contains any new points of view, but Koch's efforts to limit the cholera epidemics that in 1892/93 ravaged Europe, including Germany, and his work on the relation between bovine and human tuberculosis are treated at length. The same is true of his studies of tropical diseases, particularly malaria, and generally speaking it is obvious how Almquist concentrates on Koch's hygienic and epidemiological work. In conclusion, Almquist says the following: "Even if it is many years since Koch laid the first foundation of the modern methods of combatting epidemics and during the 1880s presented his most startling ideas, he has also after 1890 made great discoveries, which would have made anyone famous... There is so much worth rewarding in Koch's work that its very richness presents difficulties. Other scientists, nominated to the Nobel Prize, as a rule have merits that correspond to only a small fraction of what Koch has achieved, even if we only consider the period after 1890... To ignore science like this, when the Nobel Prize shall be awarded, is not possible. In all probability, the medical Nobel Committee will not for a very long time encounter a candidate, who has benefited mankind to the same extent as Koch has done."

This memo must be seen as a very strong recommendation of Koch to the Prize, in spite of the fact that the author still clings to the entirely artificial time limit of "1890" and therefore cannot give work before that time its true merit value.

For its deliberations concerning the Prize of 1905, the committee had asked Almquist to write a memo specifically about Koch's work on tuberculosis. The memo may not add very much to our picture of Koch and his achievements, but it is interesting because of the way the author organizes his material in the essay; what he attaches primary importance to. Almquist discerns what he calls "Koch's three great ideas about tuberculosis."

The first "idea," the identification and characterization of the tubercle bacillus, takes up two typewritten pages of the memo. However, the discovery itself is dealt with on just a few lines, the rest of the section is devoted to a discussion of the diagnosis, etiology and epidemiology of tuberculosis. The second "idea" concerns tuberculin and takes up one page only, with the main emphasis on its diagnostic use, at this time mainly in veterinary medicine. The third "idea," defined as the differences between the bovine and the human tubercle bacilli, takes up ten pages in the memo. It is hard to believe that these proportions really reflect Almquist's view of the relative importance of the three "great ideas." Nevertheless, that seems to be the case, judged by Almquist's account of Koch's talk at the congress on tuberculosis in 1901, where the role of cattle tuberculosis for the human disease was a main topic. Almquist must obviously have been present and he writes: "Very seldom has a scientific communication created greater sensation than Koch's talk at the tuberculosis congress in London 1901. Both his scientific conclusions and his call for changes in the fight against pulmonary tuberculosis became immediately the object of criticism and opposition. I assume that the excitement it caused is without comparison in the history of medicine, both in terms of vehemence and extent." Daring words indeed, particularly against the background of thousands of years of medical history with its innumerable great discoveries and intense conflicts. Almquist goes on to point out that now, four years later, "an immense literature has emerged as a result of Koch's short communication. Governments as well as individual scientists hastened to follow Koch's lead and investigate the problem experimentally. After four years, we can now judge this stream of publications fairly. We then find that this is a great achievement, which on an important point has meant a considerable advancement for mankind." Finally, Almquist arrives at his punch line: "It is not necessary to elaborate any further on Koch's services to mankind. The faculty of the Karolinska Institutet will very seldom have the opportunity to reward a scientific achievement, which can compare with that of Koch's work on tuberculosis."

In the year 1905 it was at long last Robert Koch's turn, and the greatest medical scientist of that period got his Nobel Prize "for his research and discoveries concerning tuberculosis." No one could possibly have any objections against this laureate. On the other hand, we can certainly argue about how the awarding institute reasoned in making its decision. It would seem that the Nobel Committee and probably also the faculty of the Karolinska Institutet had difficulties with really accepting the obvious connection, also from the point of view of merits for the Nobel Prize, between Koch's recent research and his pioneering work during the period before 1890. This is the more surprising as the bacteriology at the turn of the century was so obviously rooted in the methods that Koch had developed during the early years of his career. Instead, the committee is true to the letter of its own interpretation of the Nobel Charter in a way that sometimes produces rather remarkable arguments.

By limiting its grounds for Koch's Nobel Prize to his work on tuberculosis, the faculty and its Nobel Committee elegantly avoid discussing all his other merits that fall before the magic year of 1890. What remained was then the question of weighing the great discovery of the tubercle bacillus against his later work on tuberculin and the relation of bovine to human tuberculosis. The committee's short dismissal of the tuberculin seems reasonable. On the other hand, the number of pages that Almquist in his memo devotes to Koch's work on the relation between bovine and human tuberculosis is clearly unreasonable in view of his treatment of Koch's other achievements. The only explanation is that it represents an attempt to draw the attention to the research of the last decade in a way that is not scientifically motivated. To summarize, there can be no doubt that the recommendation of the committee and the final decision of the faculty — a Nobel Prize to Robert Koch — was well justified indeed, but that the reasoning behind the Prize seems somewhat strained.

1908

Paul Ehrlich

Nobel Committee and Nominations

The medical Nobel Committee of 1908 had the following members:

Almquist, Ernst	Professor of Hygiene
Lennmalm, Fritiof	Professor of Neurology
Mörner, Karl A. H.	Professor of Medical Biochemistry (chairman)
Pettersson, Alfred	Associate Professor of Bacteriology
Sundberg, Carl	Professor of Pathological Anatomy

Paul Ehrlich had in 1901 already been nominated once together with Behring for their work in immunology. For 1902 and 1903 he received two nominations each, 1904 nine, 1905 seven, 1906 nine and in 1907 as many as 17. Finally in 1908, he had received 12 nominations. He was thus one of the most nominated medical laureates, particularly if one considered that even after his Nobel Prize in 1908 he was nominated a further 22 more times.

Evaluations

The first evaluation of Paul Ehrlich, that dealt mainly with his immunological work, was in 1902 in a memo by Almquist and Sundberg. This "preliminary memo regarding Ehrlich's work" is thorough and penetrating compared to the rather cavalier way that the same authors treated Behring's discovery of the antibodies. Whether this reflects the authors' valuation of the two scientific personalities Behring and Ehrlich, or just indicates a more conscientious attitude to the evaluation task itself is an open question. They first gave an appreciative account of Ehrlich's immunization of animals against the plant toxins ricin and abrin. Here, the authors stressed the importance of his technique with stepwise increasing doses of toxin, which gave a better immunization effect. They also emphasized Ehrlich's demonstration of how immunity could be transmitted to the progeny, not genetically but through antibodies in the maternal milk. Almquist and Sundberg fully realized the importance of Ehrlich's distinction between passive (for instance in serum therapy) and active immunization (for instance, as a result of a certain illness, like smallpox). His methods for determining the antibody concentration in serum were hailed as being of vital importance for the successful serum therapy against diphteria. In connection with the publication of methods for the quantification of diphteria antitoxin in 1897, Ehrlich also described his side-chain theory and discussed how toxin and antitoxin (antigen and antibody) bind to each other. These questions, where at least the latter one would prove to be very controversial, had been touched on in the memo of 1902, but were further analyzed in the memos of 1903 and 1904.

The conflict with Arrhenius. In 1903, Almquist and Sundberg jointly produced a memo about Ehrlich, which like the previous one was in fact written by Sundberg, while Almquist just concurred with the opinions of his colleague. In 1904, Sundberg was the sole author of a memo about "Ehrlich's studies of the white blood cells," which concluded that this work alone was not sufficient for a Nobel Prize. In another memo, he dealt with the binding between antigen and antibody. Both Sundberg and Almquist had previously concluded that it was a question of a chemical bond and Sundberg had gone on to say: "The chemical bond between toxin and antitoxin, as well as between such bodies and antibodies in general can nowadays be considered as proven." He consequently rejected more diffuse, vitalistic explanations that had been suggested by Metchnikoff's pupil Jules Bordet, and instead agreed with Ehrlich's chemical outlook. Here he seemed to have the support of the famous Swedish physical chemist, Svante Arrhenius, whom he quoted as follows: "One of the greatest of Mr. Ehrlich's many merits is that he has always stressed that in reactions between toxins and antitoxins, as well as on

Svante Arrhenius (1859–1927).
Courtesy of the Nobel Foundation, Stockholm, Sweden.

the whole between pertinent bodies and antibodies, physical and chemical processes are taking place."

So far everything was fine between Ehrlich and Arrhenius, but the real problem was the true nature of the toxin (antigen). Was it homogenous or did it consist of disparate fractions with different properties with regard to toxicity and ability to bind antibodies? The controversy had as its starting point an experiment that Ehrlich had done where he attempted to gradually neutralize diphteria toxin by adding small portions of antitoxin. He then found that the curve obtained did not correspond to the simple one characteristic of, for instance, the neutralization (titration) of a strong acid with a strong base. It should be realized that Ehrlich always assumed that the bond between toxin and antitoxin was a strong one and that its dissociation was so small that for practical purposes it could be neglected. The reaction between toxin and antitoxin would then correspond to the titration of a strong acid by a strong base and the same type of curve would be obtained.

To explain the deviation from the expected simple titration curve, he came up with the hypothesis that the toxin was not homogenous but consisted of a mixture of what he called "toxins" (strongly toxic and strongly antibody binding); "toxons" (weakly toxic and weakly antibody binding); and "toxoids" (non-toxic but able to bind antibodies). He further assumed that this group of substances were "amboceptors", i.e. bodies that were made up of one part that could bind the antibody (the "haptophore" part) and another that was responsible for the toxicity (the "toxophore" part). The haptophore region was also supposed to participate in the binding of the different toxin fractions to the cell surface. It was this rather complicated hypothesis that Arrhenius now attacked.

Ehrlich's original method for the determination of antibody concentration was based on an animal model. What was measured was the ability of the antibody to counteract the effect of the toxin on a guinea pig weighing 250 grams under standardized conditions. This was obviously a rather complicated test system that seriously limited the number of experiments that could be performed. Ehrlich therefore worked out an alternative *in vitro* analysis that took advantage of the fact that the tetanus bacillus, in addition to its nerve toxin, also contained a poison (tetanolysin) which caused the lysis of red blood cells (hemolysis). Immunization of suitable laboratory animals with tetanolysine gave rise to an antibody against the poison (antilysin), and one now had an experimental *in vitro* system that was easy to handle and admitted of an almost unlimited number of experiments. It was this system that Arrhenius and his Danish collaborator Madsen used to investigate the interaction between an antigen and its antibody.

Ehrlich thought that the stepped curves he obtained in his experiments could be explained by his various, hypothetical types of diphteria toxins interacting differently with the antitoxin. Arrhenius and Madsen, on the other hand, observed a continuous, hyperbolic curve in their experiments, that in their opinion corresponded to the curve for the titration of a weak acid, for instance boric acid, with a weak base like ammonia. They therefore assumed that the bond between antigen and antibody could easily be dissociated and completely rejected Ehrlich's ideas of toxons and toxoids. Sundberg took care not to become involved in the controversy between the two distinguished scientists (Svante Arrhenius had been awarded the 1903 Nobel Prize in chemistry) and cautiously left the question open.

Regarding the side-chain theory, Sundberg says in his memo of 1904 that it is a valuable working hypothesis, which has very much stimulated immunological science. He summarizes his opinion of Ehrlich in the following words: "I find Ehrlich's contributions to immunology so great that he doubtless must be considered worthy of the Nobel Prize. However, in this connection, as far as his biological toxin analysis and his side-chain theory are concerned, they should so far be considered only as fruitful working hypotheses ...."

The count and professor Johansson. Count Karl A. H. Mörner was not only professor of medical biochemistry and the chairman of the medical Nobel Committee, he was also the president of the Karolinska Institutet. He was thus a more than worthy candidate to take over the task of evaluating Paul Ehrlich. Mörner had undoubtedly done a thorough job as his memo of 1906 consisted of about 70 typewritten pages. He concentrated on Ehrlich's immunological work, and on the whole considered the same problems that had already been analyzed by Almquist and Sundberg. The difference is that Mörner was much more thorough and detailed. This is particularly true of the way he discussed the interaction between antigen and antibody, a subject to which Mörner devoted a considerable part of his memo. In another memo of 1907, he returned to this question under the heading: "memo about the controversy between Arrhenius and Ehrlich" (33 pages). Finally, he took a different and more positive view of the importance of the side-chain theory as a merit for the Nobel Prize.

Mörner treaded very carefully when it came to the conflict between Arrhenius and Ehrlich. In his memo of 1907 he says: "I am reluctant to criticise Arrhenius' work in this field. My great respect for Arrhenius' contributions to science makes me unwilling to say anything that might be construed as an attempt to belittle his merits. When I feel myself compelled to go against certain statements of his, I want to restrict myself to what is absolutely essential for the question before us. I also give my comments in the form of a memo intended exclusively for the Nobel Committee, relying on that the committee will regard this memo as completely confidential. For the present it is not my intention that the memo should be communicated to the faculty."

Having thus stressed the importance of secrecy also against the faculty of the Karolinska Institutet, Mörner tried to define exactly what the conflict was all about. He found that it mainly concerned two questions:

(1) "Ehrlich is of the opinion that toxin and antitoxin, in the same way as a strong acid and a strong base, bind each other firmly and that a hydrolytic dissociation (it is not a question of an electrolytic dissociation) can for practical purposes be excluded. Arrhenius, on the other hand, maintains that the complex between antigen and antibody can be likened to the salt between a weak acid and a weak base, that in water solution is hydrolytically dissociated so that the solution contains, in addition to the salt, both free acid and free base.

The other question at issue is related to the first one.

(2) Arrhenius considers that some of the peculiarities of the reaction between toxin and antitoxin can be explained by this dissociation. Ehrlich, however, insists on his view that they depend on the toxin preparations being mixtures that contain more than one kind of antitoxin-binding substance."

An essential point for Mörner is the question of whether the reaction between antigen and antibody obeys the law of mass action and can be treated mathematically according to its principles. This was the point of view of Arrhenius, while Ehrlich denied it. Mörner spent a great deal of effort trying to show that with an alternative mathematical treatment of the data, using different principles, one could get results that are equally convincing as those obtained by Arrhenius using the law of mass action. He also sought support from the outstanding German physical chemist Walther Nernst, who according to Mörner thought that the antigen-antibody reaction was not reversible, and consequently the law of mass action did not apply. At the same time, Mörner repeatedly emphasized in his memo that the question of the true nature of the antigen-antibody reaction was not of decisive importance as far as Ehrlich's merits for the Nobel Prize were concerned. In any case, this question could not be decided until we had a much clearer picture of the structure and properties of the interacting molecules.

Regarding the side-chain theory, he was considerably more appreciative than Sundberg had been two years earlier. In his memo of 1906, Mörner says of Ehrlich's contributions to immunology: "Here his side-chain theory is particularly prominent. The experimental work of Ehrlich and his collaborators is intimately connected with this theory and its development. A possible awarding of the Nobel Prize to Ehrlich must therefore in any case include the side-chain theory as a major argument, even if the wording of the award decision were such as to attempt to disguise this fact.

Johan Erik Johansson (1862–1938).
Courtesy of the Karolinska Institutet, Stockholm, Sweden.

I do not want to conceal that I have for a long time had doubts about Ehrlich's merits for the Nobel Prize. It is in the nature of medical science that theories in medicine need a very solid experimental basis and a thorough testing before they can be considered of such significance as to merit a Nobel Prize.

After careful consideration I am of the opinion that Ehrlich's side-chain theory has arrived at such a point, thanks to the considerable experimental work that has gone into developing and defending it... I have no doubts about recommending Ehrlich to the Nobel Prize for his work in immunology, by which he has greatly broadened and deepened our knowledge of this field."

In his memo of 1907, Mörner again strongly recommended Ehrlich to the Prize and he emphasized that the controversy with Arrhenius was no real obstacle here. Perhaps it is somewhat surprising that Mörner in his arguments focuses so strongly on the side-chain theory. Undoubtedly, he appears a bit uncritical in his high appreciation of the experimental support of the theory, which at this time was hardly as strong as he maintained.

Mörner's last memo began with a pious hope that it should be kept within the Nobel Committee and not come to the attention of the faculty. This, however, proved to be an illusion. His memo was dated "July 1907," but on 30 August, Johan Erik Johansson, professor of physiology at the institute, submitted a memo of his own to the Nobel Committee, where he in a fairly uninhibited way took Mörner to task for his criticism of Arrhenius and his recommendation of Ehrlich to the Nobel Prize. Professor Johansson argued along two main lines. First, he addressed Mörner's attempt to treat the reaction between antigen and antibody mathematically, using an equation (Jellet 1875) that Johansson thought could be derived from the equation of mass action. Although he admitted that the results of Mörner's calculations agreed well enough with the values observed, he did point out that Mörner's equation required more assumptions than Arrhenius had needed. In any case, Mörner had not, according to Johansson, succeeded in showing that Ehrlich's opinion is the correct one.

Johansson's second line of argument concerned the side-chain theory which he completely rejected. He pointed out that although Ehrlich maintained that "these phenomena are of a chemical nature, in reality he has not himself understood the meaning of this assertion, nor does his theory give any opportunity of a quantitative treatment of the observations. His side-chain hypothesis can therefore hardly be called a theory. It has played the role of a working hypothesis (as pointed out in the Nobel Committee by Almquist)."

Two weeks after the committee had received Johansson's memo, Mörner responded with a "Statement at the meeting of the Nobel Committee on 14 September 1907 (not delivered or submitted in writing). Reply to the letter from professor Johansson to the committee on account of my memo regarding the controversy between Arrhenius and Ehrlich."

While Mörner had felt compelled to tread very cautiously as far as Arrhenius was concerned, he had no such qualms about Johansson. The still preserved manuscript to his "Statement" revealed the count's irritation. As usual, Mörner was fairly detailed (18 pages) as he focused on two main points. He repeated his arguments regarding the mathematical treatment of the reaction between antigen and antibody and maintained emphatically: "I fail to see that Arrhenius' categorical rejection of Ehrlich's opinion is justified." At the end of his statement, Mörner dealt with Johansson's objections against the side-chain theory:

"I now turn to the statements concerning Ehrlich's side-chain theory that professor Johansson makes on pp. 8 and 9. His account must surprise everyone who has any knowledge of this theory. The question if the binding between toxin and antitoxin follows the same principles as the reaction between a strong acid and a strong base, or alternatively between a weak acid and a weak base, is immaterial as far as the side-chain theory is concerned. The theory does not deal with this question, something that has been repeatedly pointed out in the literature. As long as professor Johansson does not explain what part of Ehrlich's side-chain theory that in his opinion is affected by the controversy between Ehrlich and Arrhenius, there is no reason or possibility to take his statement into consideration. For the present I only want to protest against professor Johansson's assertion that Ehrlich's side-chain theory has no chemical relevance. Here professor Johansson is mistaken. Ehrlich is well acquainted with modern organic chemistry and in working out his side-chain theory he has taken this knowledge as his point of departure.

What professor Johansson has said in his letter does not alter my conclusion that the controversy between Ehrlich and Arrhenius does not stand in the way of a Nobel Prize to Ehrlich."

A recommendation in writing. Ehrlich had come close to a Nobel Prize several times before, but it only finally materialized in 1908. This year it was Carl Sundberg's turn to write a memo about the candidate. He was not as wholeheartedly enthusiastic about Ehrlich as Mörner had been, particularly not where the side-chain theory was concerned:

"In what follows I am giving my opinion about the significance of this theory as a merit for the Nobel Prize and I want to emphasize the important discoveries that have been made with Ehrlich's side-chain theory as their basis. Even if the theory is not yet ready for a final judgement, the discoveries that have resulted from this hypothesis are nevertheless worthy of our recognition."

Sundberg was clearly more reserved in his judgement than Mörner and his memo did not result in a definite recommendation of Ehrlich for a Nobel Prize. Nevertheless, it would not be long before he had to take an irrevocable stand on this issue, and again it was professor Johansson who had been active. Sundberg's memo to the Nobel Committee was dated 11 August 1908, and on 15 October professor Johansson was ready with a letter to the faculty of the Karolinska Institutet: "Objections to the memo written on behalf of the Nobel Committee by professor Sundberg regarding Ehrlich 1908." The next day, he submitted some supplementary "general observations" on this subject to the faculty. Johansson repeats his previous points of view, but in these latter writings he sounded more aggressive and his disparagement of Ehrlich was more obvious. He writes, for instance, the following: "A prize to Ehrlich will absolutely, in spite of all reservations, be seen as a prize to the side-chain theory.

A question: Is it worthy of the Karolinska Institutet to give its sanction to this theory?

It has been attacked by a fellow countryman of ours. He has every chance of emerging victorious from this struggle, even if his arguments have not yet prevailed. In a question like this, where it is so difficult for most people to form an opinion of their own, one tends to agree with the opinion that has been officially sanctioned.

Should the faculty of the Karolinska Institutet intervene in this struggle with the authority of the Nobel Prize and resolutely appear as an adversary of its fellow countryman?"

Professor Johansson's letter was addressed directly to the faculty, ignoring the Nobel Committee, and this obviously annoyed the committee to such an extent that it actually resulted in something extremely unusual. Normally, the committee puts forward its suggestions *orally* to the faculty of the institute, but this time it was done in the form of a letter, dated 21 October and signed by all members of the committee.

Johansson had imprudently laid himself open to criticism by his patriotic appeal to the faculty not to side with Ehrlich in his conflict with a fellow countryman (Arrhenius) and the Nobel Committee lost no time in taking advantage of this opening. Having first dismissed Johansson's criticism of Sundberg's memo as being without interest for the committee's judgement of Ehrlich's merits, the committee nevertheless proceeded on to say the following: "Professor Johansson maintains that we should not give the Nobel Prize to Ehrlich, since he has been attacked by a fellow countryman of ours. However, it seems to us that this is going too far in patriotism, considering that the ardour with which Arrhenius has attacked Ehrlich does not essentially diminish Ehrlich's contributions to the progress of immunology."

The committee then once more reviewed Ehrlich's work in immunology, pointing out that he, among other things, had worked out the experimental methods that Arrhenius had used in his investigations of antigen-antibody interactions. According to the committee, Arrhenius' criticism of Ehrlich had not been accepted internationally. On the contrary, "the nominations of Ehrlich to the Nobel Prize have over the years shown that in the scientific community there is a strong opinion in favor of Ehrlich that has in no way diminished lately."

The committee now embarks on the unique step of submitting a recommendation *in writing* to the faculty that the Nobel Prize for 1908 should be awarded jointly to Paul Ehrlich and Elie Metchnikoff: "We maintain our suggestion that Ehrlich be awarded the Nobel Prize of this year.

We also maintain our suggestion about Metchnikoff. He has the great merit of having pioneered the modern research on immunology and to have for a long time led its development. These achievements of his have lately (through the research on opsonins etc.) been given a new vitality and importance. It would therefore not seem right to exclude him from the award.

We submit that the question of a prize to Ehrlich and Metchnikoff be solved this year, since a postponement of it would probably make it return without it having become any easier to solve.

As a reason for a prize to Ehrlich and Metchnikoff jointly, we would like to point out that in this way the faculty avoids being seen as guaranteeing the correctness of their respective theories, while at the same time recognizing the great service that they have undoubtedly both rendered to the development of immunology."

In the end, the faculty followed the advice of the committee, and professor Johansson's insinuating letter to the faculty two days later did not make a difference. In the letter, he asked: "The Nobel Committee emphasizes that the question of a prize to Ehrlich should not be postponed. But why this eagerness? Since the Nobel Committee is so sure of Ehrlich's strong position and so convinced of the unimportance of Arrhenius' criticism and likewise the lack of merit of his immunological work, why not let the matter rest so that also people outside the Nobel Committee can become equally convinced?... Could it be that the gentlemen in the Nobel Committee have a premonition that the opinion in favor of Ehrlich is decreasing? Arrhenius' work is becoming increasingly known, read and understood. Are the gentlemen afraid that Ehrlich would be a less successful candidate next year?"

Certainly professor Johansson is something of a pain in the neck for the Nobel Committee during his struggle against Ehrlich's candidacy. However, is there any substance at all in his criticism?

Undoubtedly, there are many things in Ehrlich's complicated model for the antigen-antibody reaction that can be called into question, and some of his interpretations may seem a bit strained. To that extent, at least, one can side with Arrhenius and his faithful supporter, Johansson. On the other hand, one is inclined to agree with Mörner and his colleagues that this whole question of a model for the antigen-antibody interaction is really premature, since at the time there was no structural knowledge about the interacting molecules. Furthermore, this problem is not of any real importance for Ehrlich's merits as a candidate for the Nobel Prize, which rest on a number of other contributions. The prize to Paul Ehrlich will probably always be considered as one of the most deserved of medical Nobel Prizes.

Elie Metchnikoff

Nominations

Elie Metchnikoff was nominated three times in 1901, by his faithful friend Émile Roux among others, for the discovery of phagocytosis. In 1902, 1903 and 1904, he also received three nominations and then the numbers increased steadily. At the same time, several of these nominations were marked as invalid in the protocol of the Nobel Committee. Phagocytosis and the cellular immunity had undoubtedly their fair share of advocates in the medical community.

Evaluations

The first evaluation of Metchnikoff was done in 1902 by both Almquist and Sundberg, even if it probably was written by Sundberg alone. In 1904, there appeared a new memo about Metchnikoff, this time with Sundberg as the sole author, and the same was true of 1907 and 1908. In his memo of 1904, Sundberg summarized his opinion of Metchnikoff's work: "As undeniable merits for Metchnikoff — aside from work about details — I count that he was the first to resolutely address the question of the mysterious nature of immunity. Furthermore, before anyone else he realized through his experiments that there exists a specific substance in bacteria-killing sera. This portended the discoveries of Pfeiffer and Bordet. In particular, Metchnikoff's work was an important clue for Bordet, who finally solved the problem and demonstrated that two bodies are required for bacteriolysis and hemolysis.

On the other hand, I do not think that Metchnikoff has proved his assertion of phagocytosis as the major cause of immunity, even if such cases probably can occur.

Alfred Pettersson (1867–1951).
Courtesy of the Karolinska Institutet, Stockholm, Sweden.

In a comparison between Metchnikoff and Ehrlich, the latter has undoubtedly made the greatest contribution to our knowledge."

In his memo of 1907, Sundberg drew attention to the new results by the English scientist Almrooth-Wright; the demonstration of what he called opsonins, substances that facilitated the phagocytosis of bacteria by leucocytes. Sundberg considered that the discovery of the opsonins gave a renewed interest to Metchnikoff's results, which seemed on the way to becoming overshadowed by the great progress made in antibody research. Nowadays, the opsonins are hardly believed to be of such importance. In Sundberg's opinion, a Nobel Prize in the field of immunology was not all that urgent "since it is probably desirable to wait and see how new ideas in this research will affect older ones. There is no risk in waiting since present concepts in immunology undoubtedly are in need of new and stronger arguments than have so far been forthcoming." Sundberg's memo of 1908 about Metchnikoff and immunological research was even more unconclusive. He did not even give a definitive opinion about Metchnikoff's merits or whether he was a major candidate for the Nobel Prize of 1908.

A relative newcomer in the Nobel Committee, the bacteriologist Alfred Pettersson, had also been asked to write a memo about "Metchnikoff's research concerning immunity" that same year. His report to the committee however was very different. It was not so much that he had uncovered any new facts, but unlike Sundberg he was prepared to come up with definitive recommendations. He ended his 15-page memo with the following summary of Metchnikoff's achievements: "He has advanced a theory of certain cases of immunity, based on the physiological properties of some cells (determined by numerous comparative anatomical investigations) of being able to take up and digest solid bodies intracellularly. This theory is the first that is based on real observations and has been consistently worked out.

He has carefully investigated which cells that in higher organisms function as phagocytes.

He has to a large extent been able to determine which of the different kinds of phagocytes that are active in various capacities.

He has discovered the ability of immune sera to facilitate phagocytosis.

He has demonstrated the ability of leucocytes to render bacterial toxins harmless.

He has been actively involved in the discovery of the complex nature of hemolysins and bacteriolysins.

He has worked out a new way to stimulate the activity of blood-forming organs.

Thus, it seems to me that Metchnikoff through his discoveries in immunology has shown himself worthy of being awarded the Nobel Prize."

The details of the list of merits presented by Petterson are of cause open to discussion. How much should really be credited to Metchnikoff and what is mainly the work of others? At the same time, the conclusion is undoubtedly correct. Metchnikoff's discovery of phagocytosis has had an impact on biomedical research that certainly warrants a Nobel Prize. We have already seen that this was also the decision of the faculty. Looking back over almost a century, it is obvious that the Nobel Committee of the Karolinska Institutet showed commendable foresight when it combined the name of Elie Metchnikoff with that of Paul Ehrlich.

In Defence of the Nobel Prize

In our egalitarian times, it might seem hard to justify the handing out of millions of dollars each year in prizes for something so elitistic as science and literature. Let us, nevertheless, for the sake of argument assume this to be defensible. It has after all been going on for a century by now. Another question is how the prize-awarding institutions have discharged their duties during this time. On the whole, they seem to have done fairly well, at least judging by the international prestige of the Nobel Prize. It is often said that the greatest compliment is when someone tries to imitate you. By now we already have what the media often refer to as the "alternative" Nobel Prize (The Right Livelihood Award) and the Prize in Economic Sciences given out by the Swedish National Bank in memory of Alfred Nobel — two prizes that try to borrow some of the lustre from the real Nobel Prizes.

What about Nobel's wish that the scientific prizes should be awarded to those who have done humanity the greatest service? It was, after all, what he said in his will and there is every reason to believe that he really meant it that way. For practical reasons, we have instead been obliged to award them on purely scientific grounds to whoever has done the best research. Would Nobel have been satisfied with the way we now select the scientific laureates or would he have felt that we have perverted the intentions of his will? It is of course an open question, if there is any connection at all, between scientific excellence and what benefits humanity. Have we chosen scientific criteria because they are the only ones that we can judge?

Natural science and medicine can be thought of as an ancient building of facts and theories that the scientists have built on since thousands of years. At the same time, they have labored incessantly to adapt the old structures to all the new facts of our growing knowledge. Large parts of this venerable building might seem to be completely useless, it serves no practical purposes whatsoever. If we were really motivated only by considerations of what benefits humanity, such fields of research would be a wasteland where nobody worked. Instead the scientists should crowd into fields where the usefulness of their research is particularly obvious. However, this is far from being the case. A great deal of today's research deals instead with problems where the benefit to humanity might appear, at least on the face of it, rather doubtful. It would seem to be the intellectual challenge, the unprejudiced searching for new knowledge, that more than anything else drives the scientists.

On the other hand, why should this be so reprehensible? Our curiosity about anything new is an inheritance from our primitive forefathers. Surely it is the most basic prerequisite for our development into human beings, the most curious of all apes. It is really only when it comes to the Nobel Prize that the acknowledgement of curiosity as the most fundamental motive power in research, becomes something of a moral problem. After all, there is no getting away from the fact that Nobel wanted to reward what most benefitted humanity, not just scientific excellence in general in the different fields of the Nobel Prizes. So how do we deal with this dilemma?

The only way that the prize-awarding institutions could defend their principles of selection against the possible displeasure of the great donor would be if they could establish a connection between what might, for the sake of simplicity, be called basic research and such investigations, let us call them applied research, that obviously are aimed at goals which are practically useful. This is a complicated and controversial question and perhaps we would be well advised to do what is often done in research, to look for a simpler model system. Let us disregard the humanities, with the excuse that there is after all no Nobel Prize in this field, and choose a field where the possible practical applications are obvious, for example medicine.

What is often called the medical prize is really, following the stipulations in Nobel's will, a prize in "physiology or medicine" and one might of course wonder what the donor intended here. It seems reasonable to assume that what he had in mind were, on one hand, the so-called basic sciences of medicine, i.e. physiology, biochemistry, pathology, microbiology, etc. and on the other hand, clinical medicine. In Nobel's time, physiology had long been the leading one among these basic medical sciences and it might have been close at hand to use it as a symbol for this whole group. In any case, it is a fact that a typical representative of medical basic sciences, physiology, is specifically mentioned in Nobel's will as a field for the prize. Incidentally, it might have been professor Johan Erik Johansson, whom we have already encountered as an implacable adversary of a Nobel prize to Paul Ehrlich, that originally prevailed on his friend Alfred Nobel to include a prize in physiology or medicine in his will. In all probability, Nobel must have meant that not only physiology, but basic research in general might benefit humanity. For us, who work in basic research, this is a matter of course, but among the general public, which is after all the ultimate financier of practically all basic research, this is far from being that obvious. Is there anything more, then, besides Nobel's authority, that we can base our opinion on?

To begin with, there is an obvious connection between medical progress in the form of clinically useful methods, and the general progress on all adjoining fields of research. A practically useful clinical breakthrough is inconceivable except as part of the totality of medical research; the collected knowledge that represents the present stage of medical science. Let us, however, go one step further and ask about the roots from which the individual discovery has grown. There have been rather comprehensive attempts to elucidate such scientific root systems precisely in the case of clinically important discoveries. One might have thought that such results would almost always come from applied research, clearly focused on new and clinically useful methods. Instead, they are often enough the result of discoveries that have been made in basic medical research and where the clinical usefulness has come as something of a surprise.

That would seem to be the case today, but what about the situation in the early days of the Nobel Prize? We might perhaps take the four pioneers of microbiology, who appear in this book, as a case in point. Regarding Robert Koch, it is obvious that his clinically important achievements are intimately related to his work in basic research. It was after all Koch who, to a great extent, laid the foundations of medical microbiology, worked out its basic techniques and topped off his career by identifying the tubercle bacillus as the cause of tuberculosis. His only great failure as a scientist, tuberculin, came as the result of a too ambitious project in applied research, where he attempted to find a cure for tuberculosis long before the time was ripe for this undertaking.

Of Emil von Behring, it may be said that his discovery of the antibodies resulted from work clearly aimed at a clinical goal. At the same time it is obvious how, in the same way as Koch, he fails when he devotes decades of intense research efforts to a cure for tuberculosis, something that in reality presupposed knowledge and scientific progress that was still far in the future. To a great extent it has not been attained even now, a century later.

Paul Ehrlich is something of the prototype for a scientist active in basic medical research and he has played an enormous role through his theoretical treatment of immunological problems. His brilliant therapeutic achievements with the development of salvarsan as the first great breakthrough in chemotherapy, are directly connected to his early theories about chemical affinity between dyes and cellular structures, which first led to methods for staining histological preparations and then eventually to specific drugs.

Finally, Elie Metchnikoff has himself related how, to use his own words, he started out as a marine biologist, but through the discovery of phagocytosis came to radically influence medical thinking, and ended up as a pathologist.

Perhaps we might hopefully claim to work according to the intentions of the donor when we select the scientific Nobel laureates using criteria that only take the quality of their research into account, disregarding the question of how immediately useful the results might be. There is, after all, much to indicate that even the most esoteric basic research may sometimes turn out to be of the greatest benefit to humanity.

Bibliography

Behring, Emil. "Über Iodoform und Iodoformwirkung." *Deutsche med. Wochenschrift* **8**, 146–148 (1882).

Behring, Emil and Kitasato, Shibasaburo. "Über das Zustandekommen der Diphterie-Immunität und der Tetanus-Immunität bei Tieren." *Deutsche med. Wochenschrift* **16**, 113–114 (1890).

Behring, Emil. "Untersuchungen über das Zustandekommen der Diphterie-Immunität bei Tieren." *Deutsche med. Wochenschrift* **16**, 1145–1148 (1890).

Behring, Emil. *Tuberkulosebekämpfung* (Marburg, 1903).

Behring, Emil. "Über ein neues Diphterieschutzmittel." *Deutsche med. Wochenschrift* **39**, 873–876 (1913).

Ehrlich, Paul. *Beiträge zur Theorie und Praxis der histologischen Färbung* (Inaug. dissert., Leipzig University, 1878).

Ehrlich, Paul. "Experimentelle Untersuchungen über Immunität. I. Über Ricin. II. Über Abrin." *Deutsche med. Wochenschrift* **17**, 976–979, 1218–1219 (1891).

Ehrlich, Paul. "Die Wertbemessung des Diphterieheilserums und deren theoretische Grundlagen." *Klinische Jahrbuch* **6**, 299–326 (1897).

Ehrlich, Paul. "On Immunity, With Special Reference to Cell Life." *Proc. Royal Soc.* **66**, 424–448 (1900).

Ehrlich, Paul. "Die Behandlung der Syphilis mit dem Ehrlichschen Präparat 606." *Deutsche med. Wochenschrift* **36**, 1893–1896 (1910).

Ehrlich, Paul. "Die Salvarsantherapie. Rückblicke und Ausblicke." *Münchener med. Wochenschrift* **58**, 1–10 (1911).

Greuling, Walter. *Paul Ehrlich. Leben und Werk* (Econ-Verlag, Düsseldorf, 1954).

Kathe, Johannes. *Robert Koch und sein Werk* (Akademie Verlag, Berlin, 1961).

Koch, Robert. "Die Aetiologie der Milzbrand-Krankheit, begründet auf die Entwicklungsgeschichte des Bacillus Antracis." *Beiträge zur Biologie der Pflanzen* **2**, 277–311 (1876).

Koch, Robert. "Verfahren zur Untersuchung, zum konservieren und photographieren der Bakterien." *Beiträge zur Biologie der Pflanzen* **2**, 399–434 (1877).

Koch, Robert. "Zur Untersuchung von pathogenen Organismen." *Mitteilung aus dem Kaiserlichen Gesundheitsamt* **1**, 1–48 (1881).

Koch, Robert. "Die Aetiologie der Tuberculose." *Berliner klinische Wochenschrift* **19**, 221–230 (1882).

Koch, Robert. "Weitere Mitteilungen über ein Heilmittel gegen Tuberculose." *Deutsche med. Wochenschrift* **16**, 1029–1032 (1890).

Koch, Robert. "Cholera-Berichte aus Egypten und Indien." *Deutsche Vierteljahrsschrift für öffentliche Gesundheitspflege* **16**, 493–515 (1884).

Koch, Robert. "The Fight Against Tuberculosis in the Light of Experience Gained in the Successful Combat of Other Infectious Diseases." *Br. Med. J.* **2**, 189–193 (1901).

Metchnikoff, Elie. *Lecons sur la pathologie comparée de l'inflammation* (Institut Pasteur, 1892).

Metchnikoff, Elie. *L'immunité dans les maladies infectieuses* (Institut Pasteur, 1901).

Metchnikoff, Olga. *Life of Elie Metchnikoff* (Constable & Co, London, 1921).

Möllers, Bernhard. *Robert Koch. Persönlichkeit und Lebenswerk* (Schmorl und von Seefeld Nachf., Hannover, 1950).

Schück, Henrik; Sohlman, Ragnar; Österling, Anders; Liljestrand, Göran; Westgren, Arne; Siegbahn, Manne; and Schou, August. *Nobel, The Man and His Prizes* (The Nobel Foundation, Ed., Stockholm, 1950).

Zeiss, H. and Bieling, R. *Behring — Gestalt und Werk* (Bruno Schultz Verlag, Berlin-Grunewald, 1940).

Index